Safeguarding Satellite Communications

IEEE Press

445 Hoes Lane

Piscataway, NJ 08854

Safeguarding Satellite Communications

Issues, Challenges, and Solutions

Jianping An

Beijing Institute of Technology
Beijing, China

Shuai Wang

Beijing Institute of Technology
Beijing, China

Pingyue Yue

Beijing Institute of Technology
Beijing, China

Gaofeng Pan

Beijing Institute of Technology
Beijing, China

IEEE Press
WILEY

Published by John Wiley & Sons, Inc., Hoboken, New Jersey.

Published simultaneously in Canada.

Library of Congress Cataloging-in-Publication Data has been applied for:

Hardback ISBN: 9781394304295

ePDF ISBN: 9781394304318

ePub ISBN: 9781394304301

oBook ISBN: 9781394304325

Cover Design: Wiley
Cover Image: © Andrei Armiagov/Shutterstock

To our families!

Contents

Foreword

In today's digital age, satellite communication has become a critical infrastructure that supports global information exchange. Its applications are widespread across various crucial fields, including military defense, commercial operations, scientific research, and public welfare. From strategic communications and intelligence gathering in the military domain to broadcasting, television, and internet access in the commercial sector to data transmission in scientific exploration and emergency communication support during disaster relief, satellite communication is omnipresent, providing indispensable support for the efficient functioning of modern society.

However, as the applications of satellite communication systems continue to expand and deepen, security concerns have increasingly become a critical issue, limiting their further development. The openness and broadcast nature of satellite communication make it particularly vulnerable to various attacks, such as malicious interference, eavesdropping, and forgery. These attacks can not only lead to communication disruptions and information leaks but also trigger severe security incidents, posing potential threats to national sovereignty, economic development, and social stability. Therefore, in-depth research into satellite communication security technologies to enhance system security and reliability has become an urgent priority.

This book focuses on the important issue of satellite communication security, providing a comprehensive, in-depth, and systematic analysis of the security challenges faced by satellite communication systems. Beginning with the development history and system architecture of satellite communication, it gradually explores the nature, types, and corresponding countermeasures of security issues. The content covers key areas such as security risks, security requirements, security enhancement solutions, design standards, and future development directions.

This comprehensive approach constructs a complete knowledge system for readers, helping them gain a deeper understanding of the core issues and cutting-edge technologies in the field of satellite communication security.

The book not only provides a detailed explanation of traditional security protection technologies but also delves into the application of emerging technologies, such as quantum communication, blockchain, terahertz communication, and artificial intelligence, in satellite communication security. These innovative technologies offer new ideas and solutions for addressing the security challenges of satellite communication, demonstrating the significant growth potential of this field. Additionally, the book offers a detailed analysis of the security performance of satellite communication systems in different orbital environments and proposes practical solutions to the unique challenges faced by low Earth orbit satellite communication systems, with strong relevance and practicality.

As a professional working in the field of satellite communication security, this book holds significant academic value and practical guidance. It not only provides researchers in related fields with a wealth of theoretical knowledge and research insights, laying a solid foundation for further academic exploration, but also offers valuable practical guidance for engineers and technicians involved in the design, operation, and management of satellite communication systems. This guidance will help them better address various security challenges in real-world applications and enhance the security and reliability of these systems.

It is believed that the publication of this book will play an active role in advancing the development of satellite communication security, promoting academic research and technological innovation in this field, and providing strong support for ensuring the safe and stable operation of satellite communication systems.

Nanyang Technological University, Singapore *Dusit (Tao) Niyato*
March 20, 2025

Preface

In the contemporary context of global interconnectivity, satellite communication has emerged as an indispensable conduit linking every region of the planet. Its utility is profoundly integrated into diverse facets of society, encompassing routine communication and entertainment, while also bolstering essential infrastructure pivotal to national economies and livelihoods, in addition to facilitating advancements in scientific inquiry related to space exploration. The importance of satellite communication is becoming increasingly conspicuous. Nonetheless, as the magnitude of satellite communication systems continues to escalate and the contexts of their application become more intricate, security concerns have gradually surfaced as the principal obstacle hindering their progress. These security vulnerabilities not only possess the capacity to engender information breaches and communication interruptions but may also detrimentally affect national security, economic advancement, and societal stability. It is with a substantial comprehension of the prevailing circumstances that we have resolved to author this book with the objective of delivering a thorough and extensive examination within the discipline of satellite communication security, thereby facilitating the progression of this significant field.

Throughout the composition of this manuscript, we have endeavored to comprehensively address all pivotal dimensions of satellite communication security. Initiating with the fundamental architecture of satellite communication systems, we provide an exhaustive examination of the components and functionalities of the space segment, ground segment, and user segment, thereby affording readers a lucid comprehension of the holistic system. Subsequently, we investigate the security dilemmas confronting satellite communication, encompassing an array of attack methodologies, reliability vulnerabilities, and associated security mandates, which empowers readers to grasp the intricate and multifaceted nature of these security concerns. In response to these challenges, we proffer a spectrum of security enhancement strategies, integrating avant-garde technologies such as quantum communication and blockchain, alongside traditional methodologies such as interference coordination and spread spectrum technologies, thus presenting a wealth of concepts and techniques for addressing pragmatic security issues. Furthermore, we furnish a thorough analysis of the design standards pertinent to satellite communication systems, encompassing elements such as orbit selection, frequency allocation, and payload design, thereby offering both theoretical underpinnings and pragmatic counsel for assuring the security and dependability of system architecture. Finally, we investigate the prospective developmental trajectories of satellite communication security technologies and deliberate on the potential applications of emergent technologies within this domain, thereby prompting readers to contemplate future research avenues.

We aspire for this volume to emerge as a pivotal reference within the domain of satellite communication security, providing a thorough theoretical framework alongside an extensive array of research propositions for scholars engaged in pertinent investigations, thereby facilitating their attainment of novel advancements in their scholarly pursuits. Concurrently, it will function as a pragmatic technical manual for engineers and technicians involved in the design, implementation, and operation of satellite communication systems, empowering them to more effectively confront security challenges and enhance the security and reliability of satellite communication infrastructures.

Despite the substantial effort invested in the composition of this text, the field of satellite communication security is undergoing rapid evolution, characterized by the incessant emergence of new technologies and challenges. We recognize that this book may possess certain constraints, and we eagerly anticipate receiving constructive feedback and recommendations from readers, which will assist us in perpetually enhancing and refining our research endeavors in the future. We aspire for this book to actively contribute to the advancement of satellite communication security and collectively aid in the establishment of a more secure and dependable satellite communication network.

In conclusion, we wish to convey our sincere appreciation to Ziyi Yang for vital contributions to chapter structure; Xuyang Zhang and Zhen Liu for their invaluable contributions to Chapters 1, 2, and 4; to Hang Deng, Yanci Si, Jia Ye, and Zihan Ni for their significant input in Chapters 3, 5, 6, 7, and 8; and to Yanxin Wu, Zhen Liu, Yanci Si, Yi Tao, and Haoxing Zhang for their essential contributions to Chapters 9, 10, 11, 12, and 13. Furthermore, we extend our gratitude to Yi Tao and Xiaorui Zhang for the exceptional illustrations they crafted for this volume.

Jianping An, Shuai Wang, Pingyue Yue, and Gaofeng Pan

Beijing, China

August 8, 2025

About the Authors

Jianping An received the PhD degree from the Beijing Institute of Technology, China, in 1996. He joined the School of Information and Electronics at the same institution in 1995 and is currently a full professor there. He is currently Dean of the School of Cyberspace Science and Technology at the Beijing Institute of Technology. His research interests include digital signal processing, wireless communications and satellite networks. He has received two national awards for technological inventions and scientific and technological progress.

Shuai Wang received the PhD degree in communications systems from the Beijing Institute of Technology (BIT), China, in 2012. Upon graduating, he joined the faculty of the School of Information and Electronics at BIT. In 2021, he transferred to the newly founded School of Cyberspace Science and Technology, where he was appointed Chair Professor of the Department of Information Security and Countermeasures. He has published over 40 peer-reviewed articles in leading IEEE journals and conference proceedings and holds over 60 patents. His research interests include satellite communications, anti-interference communications, and datalink technologies for space platforms. In 2019, he was a co-recipient of the Second Class National Technical Invention Award of China. He has served as an editor for IEEE Communications Letters and IEEE Wireless Communications Letters. He currently serves as an editor for China Communications.

Pingyue Yue received the BSc degree from Zhengzhou University, Zhengzhou, China, in 2016 and the PhD degree from the Beijing Institute of Technology in 2024. He is currently a postdoc at the School of Cyberspace Science and Technology at the Beijing Institute of Technology in China. His research interests include satellite communication, physical-layer security, and interference suppression.

Gaofeng Pan received the BSc degree in Communication Engineering from Zhengzhou University in Zhengzhou, China, in 2005, and the PhD degree in Communication and Information Systems from Southwest Jiaotong University in Chengdu, China, in 2011. He is currently a Professor at the School of Cyberspace Science and Technology at the Beijing Institute of Technology in China. His research interests include special topics in communications theory, signal processing, and protocol design. He also serves as an editor for several journals, including IEEE Transactions on Communications and IEEE Transactions on Green Communications and Networking.

Acknowledgments

First, we would like to thank the reviewers and editors at Wiley—Sandra Grayson, Becky Cowan, Kavipriya Ramachandran, and Veena Rajendran—whose careful reviews and valuable feedback have helped to improve this book.

We are also grateful to the National Natural Science Foundation of China (NSFC) for providing us with several research grants. In particular, the grant for investigating various types of satellite and secure communications resulted in high-quality journal papers that were shared with researchers. Encouraged by their enthusiastic response to our work, we published the tutorial paper "Low Earth Orbit Satellite Security and Reliability: Issues, Solutions, and the Road Ahead" in IEEE Communications Surveys & Tutorials (Third Quarter, 2023). Furthermore, encouraged by the positive response to the tutorial paper—namely that it was selected as an ESI highly cited paper in March/April 2025—this hardens our idea of writing this book detailing the security challenges faced by satellite communication systems.

We would like to thank our collaborators on satellite and secure communications in recent years: Mohamed-Slim Alouni at King Abdullah University of Science and Technology (KAUST), Saudi Arabia; Lajos Hanzo at the School of Electronics and Computer Science, University of Southampton, UK; John Thompson at the University of Edinburgh, UK; Halim Yanikomeroglu at Carleton University, Canada; Pei Xiao at the Institute for Communication Systems, the University of Surrey, UK; Jia Ye at the School of Electrical Engineering, Chongqing University, China; and Jiankang Zhang at the Department of Computing and Informatics, Bournemouth University, UK.

We are very grateful for the support we received from Dusit (Tao) Niyato at Nanyang Technological University in Singapore.

We would also like to thank our graduate students for their support and suggestions over the past few years.

Finally, we would also like to express our deep gratitude to our families for their support and help with the book.

Jianping An, Shuai Wang, Pingyue Yue, and Gaofeng Pan.

Acronyms

3GPP	3rd generation partnership project
5G	5th generation mobile networks
6G	6th generation mobile networks
ACM	Adaptive coding and modulation
ACS	Attitude control systems
ADC	Analog-to-digital converter
AES	Advanced encryption standard
AFMS	Autonomous fault management system
AI	Artificial intelligence
AMC	Adaptive modulation and coding
AN	Artificial noise
APC	Adaptive power control
ARIMA	Autoregressive integrated moving average
ASC	Average secrecy capacity
ASIC	Application-specific integrated circuit
AWGN	Additive white Gaussian noise
BEC	Binary error channel
BER	Bit error rate
BF	Beamforming
BLER	Blocklength error rate

BP	Belief propagation
BPSK	Binary phase shift keying
CCI	Co-channel interference
CCSDS	Consultative committee for space data systems
CDF	Cumulative distribution function
CDMA	Code division multiple access
CIR	Channel impulse response
CNN	Convolutional neural networks
COTS	Commercial off-the-shelf
CPU	Central processing unit
CRC	Cyclic redundancy check
CSI	Channel state information
CV	Computer vision
DDoS	Distributed denial of service
DL	Deep learning
DoS	Denial of service
DSP	Digital signal processing
DSSS	Direct-sequence spread spectrum
DTH	Direct-to-home
DVB	Digital video broadcasting
DVB-RCS	Digital video broadcasting-return channel via satellite
E2EE	End-to-end encryption
ECC	Error correction code
ECSS	European Cooperation for Space Standardization
ELM	Extreme learning machine
ESA	European Space Agency
ETSI	European Telecommunications Standards Institute
FBL	Finite blocklength
FBMC	Filter-bank multi-carrier
FDA	Frequency diverse array
FDM	Frequency division multiplexing

FDMA	Frequency division multiple access
FEC	Forward error correction
FFHSS	Fast frequency hopping spread spectrum
FH	Frequency hopping
FHSS	Frequency hopping spread spectrum
FMECA	Failure modes, effects, and criticality analysis
FPGA	Field programmable gate array
FSO	Free space optical communication
FTA	Fault tree analysis
GEO	Geostationary Earth orbit
GPS	Global positioning system
GPU	Graphic processing unit
GRU	Gated recurrent unit
GTO	Geostationary transfer orbit
HAP	High altitude platform
HARQ	Hybrid automatic repeat request
HEO	Highly elliptical orbit
HTS	High-throughput satellite
ICI	Inter-carrier interference
IDS	Intrusion detection system
INTELSAT	International Telecommunications Satellite Organization
IoT	Internet of Things
IoRT	Internet of remote things
ISAC	Integrated sensing and communication
ISAR	Inverse Synthetic Aperture Radar
ISFFT	Inverse symplectic finite Fourier transform
ISL	Inter-satellite link
ISR	Interference-to-signal ratio
ITU	International Telecommunication Union
LDPC	Low-density parity-check
LEO	Low Earth orbit
LFDA	Linear frequency diverse array

LMS	Least mean square
LoS	Line of sight
LSTM	Long short term memory
MAI	Multiple access interference
MBSE	Model-based system engineering
MEO	Medium Earth orbit
MFA	Multi-factor authentication
MIMO	Multiple-input multiple-output
MITM	Man-in-the-middle
ML	Machine learning
MLI	Multilayer insulation
mMTC	massive machine type communication
mmWave	Millimeter wave
MSL	Mars Science Laboratory
NASA	National Aeronautics and Space Administration
NCC	Network control center
NFV	Network function virtualization
NR	New radio
NTN	Non-terrestrial network
OBP	On-board processing
OFDM	Orthogonal frequency division multiplexing
OFDMA	Orthogonal frequency division multiple access
OGS	Optical ground station
OISL	Optical inter-satellite link
OP	Outage probability
OTFS	Orthogonal time frequency space
PA	Phased array
PAPR	Peak-to-average power ratio
PDF	Probability density function
PER	Packet error rate
PG	Processing gain
PIC	Photonic integrated circuit
PLA	Physical layer authentication
PLE	Physical layer encryption

PLKG	Physical layer key generation
PLL	Phase-locked loop
PLS	Physical layer security
PO	Polar orbit
PPC	Pay-per-click
PQC	Post-quantum cryptography
PR	Pseudo-random
PRN	Pseudo-random noise
PSD	Power spectral density
PU	Primary user
QAM	Quadrature amplitude modulation
QKD	Quantum key distribution
QoS	Quality of service
QPSK	Quadrature phase shift keying
QSDC	Quantum secure direct communication
RAID	Redundant arrays of inexpensive disks
RAM	Random access memory
RAN	Radio access network
RBAC	Role-based access control
R-CNN	Region-based CNN
RF	Radio frequency
RFID	Radio frequency identification
RIS	Reconfigurable intelligent surface
RL	Reinforcement learning
RM	Reed–muller
RNN	Recurrent neural network
RS	Reed–solomon
RSO	Resident space object
SA	Service aspects
SBR	Spaceborne radar
SBC	Spaceborne camera
SC	Successive cancellation
SCC	Sensing, communication and computation

SC-FDMA	Single carrier frequency division multiple access
SCMA	Sparse code multiple access
SCPS	Space communication protocol specification
SCS	Satellite communication system
SDN	Software defined networking
SDR	Software-defined radio
SEE	Single event effect
SEU	Single event upset
SFFT	Symplectic finite Fourier transform
SINR	Signal-to-interference-plus-noise ratio
SNR	Signal-to-noise ratio
SOI	Silicon-on-insulator
SOP	Secrecy outage probability
SRAM	Static random-access memories
SS	Spread spectrum
SSA	Space situational awareness
SSO	Sun-Synchronous orbit
SSPA	Solid-state power amplifier
STI	Shallow trench isolation
SU	Secondary user
TDM	Time division multiplexing
TDMA	Time division multiple access
THz	Terahertz
TID	Total ionizing dose
TMR	Triple module redundancy
TOA	Time-of-arrival
TPC	Transmit precoding
TT&C	Telemetry, tracking and control
TWTA	Traveling wave tube amplifier
UAV	Unmanned aerial vehicle
UE	User equipment
UNOOSA	United Nations Office for Outer Space Affairs

URLLC	Ultra-reliable low-latency communication
V2X	Vehicle-to-everything
ViT	Vision transformer
VLEO	Very low earth orbit
WLAN	Wireless local Area network
YOLO	You only look once

Introduction

This volume serves as a scholarly monograph concerning the security and efficiency technologies pertinent to satellite communication systems. In the contemporary digital landscape, satellite communication plays an indispensable role in facilitating global information dissemination; however, the associated security dilemmas are increasingly profound. The current book seeks to elucidate these challenges.

The text initiates with a comprehensive overview of the historical development of satellite communication, emphasizing pivotal milestones from its genesis to its technological evolution. Subsequently, it delves into the system architecture, encompassing the space, ground, and user segments while also addressing the myriad challenges encountered by each segment. Regarding security, the book delineates various types of attacks, elucidating their foundational principles and inherent risks in conjunction with the requisite security requirements.

In response to these adversities, the text proffers solutions, such as quantum communication and blockchain technology, for proactive security measures alongside advanced antennas and interference coordination techniques for passive security enhancements. Additionally, it examines satellite orbits, frequency selection, payload configuration, and the principles of physical layer security.

A salient feature of this book is its adept integration of theoretical frameworks with pragmatic applications. It renders significant insights for both researchers and engineers, thereby fostering the progression of security technologies within the realm of satellite communication.

Part I

Fundamentals

1

Introduction

As an exceptional amalgamation of scientific principles and technological advancements, satellite communication represents a complex system that facilitates interactions between the terrestrial and celestial spheres. This system is meticulously engineered to traverse extensive distances, thereby allowing for seamless real-time communication across nations, seas, and isolated regions. The basic principle is the transmission of data through electromagnetic waves, which are dispatched from a ground station, transmitted via a satellite positioned in orbit, and subsequently received by another terrestrial station. The synchronized interaction between ground-based infrastructure and space-based technology forms the essential paradigm of worldwide communication [1].

The fundamental tenet of satellite communication encompasses the propagation of electromagnetic waves, thereby enabling extensive geographical connectivity. A ground station transmits signals toward orbiting satellites utilizing high-frequency carriers, wherein the uplink procedure necessitates precise calibration of signal intensity and directional alignment to effectively reach the satellite's communication payload [2]. Upon receipt, the satellite's transponder amplifies and reconfigures the received signal via modulation, frequently employing frequency conversion techniques to mitigate interference from concurrent transmissions. The processed signal is subsequently prepared for downlink transmission to guarantee an undistorted return of information to the terrestrial domain. Following this procedure, the satellite dispatches the modified signal toward specified terrestrial receivers. These signals are subsequently decoded by ground stations or terminal devices into their original informational formats, which may include voice communications, video transmissions, or Internet data packets. The downlink mechanism underscores the importance of transmission reliability, notwithstanding the inherent challenges posed by planetary-scale propagation phenomena.

Safeguarding Satellite Communications: Issues, Challenges, and Solutions, First Edition.
Jianping An, et al.

Typically, satellite communications have revolutionized television broadcasting by facilitating the global distribution of signals, particularly benefiting regions that lack robust terrestrial communication infrastructures. This technological advancement guarantees that remote locales receive high-quality programming and additionally functions as an essential component of infrastructure during natural disasters and humanitarian emergencies. In scenarios where terrestrial systems become nonoperational, the utilization of satellites ensures the persistence of dependable communication channels, thereby enabling a multitude of critical operations. These operations encompass emergency response coordination, the swift deployment of relief services, and the dissemination of essential information crucial for survival [3].

1.1 A Brief History of Satellite Communications

In the era characterized by an unprecedented proliferation of information, the radio waves traveling over the satellites silently announce the story of human wisdom. Communication satellites function as "messengers" in the sky, facilitating the transmission of information across vast distances on the terrestrial globe. The inception of satellite communication stands as one of the extraordinary accomplishments of contemporary science and technology. From the nascent idea to the advanced technological iterations, satellite communication has not only transformed the modalities of human interaction but has also exerted a substantial influence on socioeconomic progress. Figure 1.1 delineates a chronological depiction of the evolution of satellite communication, highlighting the motivations for its inception, the contextual background, and significant milestones [4].

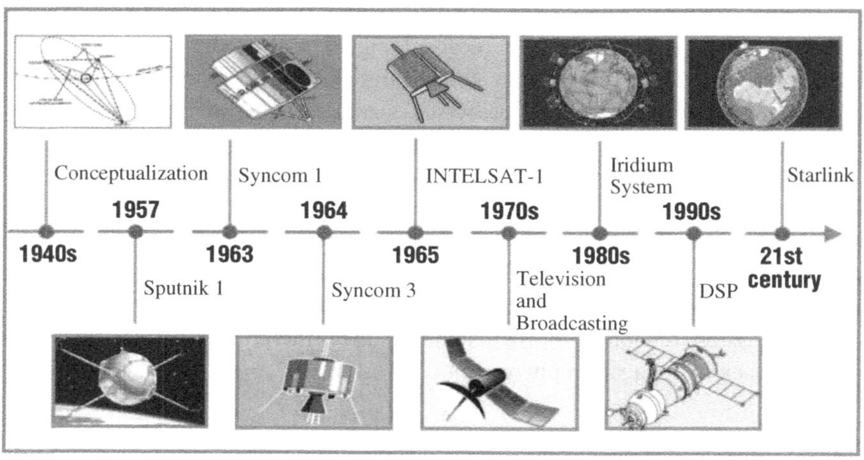

Figure 1.1 History of satellite communications.

The Conceptualization: Arthur C. Clarke and the idea of Geostationary Satellites in 1945. During World War II, advancements in global communication technologies encountered formidable obstacles. Conventional radio wave communication was constrained by the Earth's curvature, thereby limiting the transmission range of signals. Long-distance communication, particularly over oceanic expanses, necessitated intricate relay stations or the deployment of undersea cables. The exigencies of military operations during the conflict motivated scientists to investigate innovative solutions aimed at surmounting the limitations associated with long-distance communication. The British scientist and author Arthur C. Clarke articulated a groundbreaking concept in his 1945 paper published in Wireless World, titled "Extra-Terrestrial Relays," wherein he proposed the implementation of geostationary satellites to facilitate global telecommunications. His transformative proposition delineated an orbital altitude of 35,786 km, at which satellites would exhibit synchronous rotation with Earth, thereby enabling continuous signal coverage over specified terrestrial areas through fixed orbital trajectories. Clarke's foresight established the theoretical groundwork for the development of contemporary communication satellites. Despite the visionary nature of Clarke's proposal, the technological infrastructure of the era was inadequate to actualize the concept of geostationary satellites. The domain of rocket technology was nascent, and the precise electronic apparatus requisite for satellite operations had yet to be developed. Consequently, Clarke's concept remained within the realm of theory, as the necessary technological capabilities to actualize it were unavailable.

The inaugural artificial satellite, Soviet Sputnik 1, was successfully launched in 1957. During the Cold War, the domain of space exploration emerged as a pivotal battleground for the United States and the Soviet Union to assert their supremacy on a global scale. The Soviet Union sought to exhibit its scientific prowess through the launch of the inaugural artificial satellite, which concurrently served to evaluate the reliability of rocket technology crucial for subsequent advancements. On 4 October 1957, the Soviet Union ushered in the space age by achieving a significant historical milestone: the deployment of Sputnik 1, recognized as humanity's first artificial satellite orbiting Earth. With a mass of 83.6 kg, it functioned within low Earth orbit (LEO) and transmitted fundamental radio signals. Although it was not specifically engineered for communication purposes, it validated the concept that terrestrial objects could be successfully placed in space, thereby igniting a worldwide fascination with space exploration. Notwithstanding its achievements, Sputnik 1 exhibited several significant limitations. It facilitated only rudimentary scientific experiments, experienced constrained coverage attributable to its low orbital altitude, and endured for merely a few weeks before disintegrating upon reentry. These constraints underscored the imperative for subsequent technological advancements.

Geostationary satellites, heralding the advent of the Communication Era, commenced their deployment in 1963. Following the Soviet Union's successful launch of Sputnik 1, the United States expedited its space exploration initiatives to attain technological supremacy. Geostationary satellites, which are proficient in providing fixed coverage over designated regions, emerged as the optimal solution for facilitating global communication. The United States successfully launched Syncom 1, marking the inception of geostationary satellites. Despite its experimental nature, this satellite demonstrated the viability of geostationary orbits, establishing a foundation for subsequent communication satellites. The ensuing launches facilitated the transition of satellite communication from abstract theory to tangible application. Initial geostationary satellites encountered several challenges, including limited signal capacity, significant latency attributable to orbital distance, and elevated costs. These constraints underscored the imperative for continued technological advancement.

Syncom 3 and the live transmission of the Tokyo Olympics in 1964. The 1964 Tokyo Olympics offered Japan a platform to exhibit its economic resurgence and technological sophistication to a global audience. In order to facilitate a worldwide live transmission of the event, Japan opted to employ satellite communication technology, thereby surmounting the constraints associated with conventional television signal transmission. Japan adeptly launched the Syncom 3 communication satellite. For the inaugural instance, a worldwide live transmission of the Tokyo Olympics was accomplished via satellite technology. This event signified the initial application of satellite communication within the context of a significant international occasion. It not only illustrated the operational utility of satellite communication but also expedited the commercialization of satellite technology. Nonetheless, Syncom 3 possessed certain deficiencies. Its coverage area was limited, and the quality of the broadcast was constrained by the technological capabilities prevalent at the time. Furthermore, the transmission of live content still depended on terrestrial infrastructure, and the overall expenditure remained substantial.

The inaugural commercial communication satellite, known as the International Telecommunications Satellite Organization (INTELSAT), was successfully launched in the year 1965. The escalating necessity for international communication, particularly in terms of transoceanic linkages, underscored the inadequacies of conventional submarine cables, which were incapable of facilitating real-time voice and video interactions. In response to this predicament, the INTELSAT organization was founded in 1964. The primary objective of this initiative was to orchestrate global resources for the advancement of communication satellites. INTELSAT achieved a significant milestone by launching the world's first commercial communication satellite, INTELSAT-1, colloquially referred to as "Early Bird." This pioneering satellite enabled the transmission of telephone, telegraph, and television signals, thereby signifying the initial achievement of real-time intercontinental communication. INTELSAT-1 established the groundwork for the commercialization of satellite communication and considerably enhanced global information dissemination. Notwithstanding its revolutionary contributions, INTELSAT-1 exhibited significant limitations. Its capacity was confined to approximately 240 telephone circuits or a singular television channel. Furthermore, the quality of communication was susceptible to atmospheric disruptions and contingent upon the efficacy of ground stations. These constraints emphasized the necessity for continued advancements in satellite technology.

The Proliferation of Satellite Communication: The pinnacle era of television and broadcasting in the 1970s. The extensive integration of television and radio on a global scale engendered an escalating demand for large-scale signal transmission. Due to its expansive coverage and instantaneous capabilities, satellite communication emerged as the fundamental technology for disseminating television and radio signals. Numerous nations have initiated the deployment of communication satellites explicitly designated for broadcasting and television relay functions. Satellites during this epoch were predominantly employed to convey international news, sporting events, and entertainment programs, thus signifying a golden age for satellite communication within the media landscape. Furthermore, the exorbitant costs associated with satellite construction and launch constituted considerable financial impediments, thereby limiting the involvement of smaller nations and developing countries. This situation underscored the inequitable accessibility of satellite technology during this time.

Motorola's Iridium System: An endeavor in global mobile communication initiated in the 1980s [5]. The elevated latency and restricted coverage associated with geostationary satellites were inadequate to satisfy the escalating requirements of mobile communication. To mitigate these obstacles, Motorola proposed the development of an LEO satellite constellation to enable comprehensive global coverage. The Motorola Iridium system comprised 66 LEO satellites, specifically engineered to deliver global voice and data communication services. Its low-orbit architecture substantially diminished communication latency while guaranteeing worldwide coverage, encompassing both remote regions and maritime areas. This pioneering methodology rendered it one of the most ambitious satellite communication initiatives of its era. However, the Iridium system encountered exorbitantly high construction and maintenance expenditures, coupled with market demand that fell short of projections, ultimately culminating in its commercial demise. This case underscored the critical necessity of aligning technological advancements with pragmatic market requisites.

Technological Advancements: Innovations in the digital era, during the 1990s. The escalating demand for enhanced bandwidth and improved efficiency in communication illuminated the constraints inherent in traditional satellite systems. Progress in digital signal processing (DSP) presents new opportunities for addressing these obstacles. In that era, satellite communication integrated technologies such as digital modulation, data compression, and multi-beam antennas. These advancements significantly improved both efficiency and capacity. For example, multi-beam antennas enabled satellites to serve multiple regions, thereby meeting varied market demands simultaneously. Although these developments represented considerable advancements, satellite communication continued to fall short of terrestrial fiber-optic systems in terms of cost-effectiveness and overall efficiency. This disparity underscored the ongoing necessity for innovation to maintain a competitive edge.

The Internet Epoch: High-throughput satellites (HTS) in the 21st century. The unprecedented expansion of the Internet has substantially elevated the global demand for broadband services. This necessity was particularly pronounced in remote locales, aboard maritime vessels, and within aerial frameworks, where conventional satellites encountered significant challenges in providing the requisite data transmission speeds. The HTS emerged as a viable solution, utilizing multi-beam technologies and frequency reuse methodologies to enhance capacity markedly. For instance, ViaSat's HTS systems accomplished data transmission velocities reaching hundreds of gigabits per second, thereby facilitating applications such as in-flight broadband connectivity and distance learning initiatives. Despite the considerable advancements offered by HTS, the exorbitant costs associated with deployment and protracted implementation timelines persist as formidable obstacles. Moreover, the advent of LEO constellations, exemplified by Starlink, has introduced formidable competition, thereby undermining the preeminence of traditional HTS systems.

1.2 Development of Satellite Communications

1.2.1 Classification of Satellite Orbits

In the majority of instances, satellites are deployed into a predefined orbital path surrounding Earth subsequent to their launch. Nonetheless, in certain exceptional scenarios, a satellite may undergo a process of orbital realignment, resulting in its traversal along a modified orbital trajectory. Satellites are typically categorized based on their orbital altitude (i.e., the distance from the Earth's surface), which significantly influences their coverage area and velocity as they orbit Earth. When selecting the appropriate type of orbit, developers must take into account their intended purposes, the data to be collected, the services to be rendered, as well as the associated costs, coverage regions, and the practicality of various orbital configurations. According to their respective orbits, satellites are classified into five principal types: LEO, medium Earth orbit (MEO), geostationary orbit (GEO), sun-synchronous orbit (SSO), and geostationary transfer orbit (GTO), as illustrated in Figure 1.2.

As illustrated in Table 1.1, LEO satellites operate at altitudes ranging from approximately 500–2,000 km above the terrestrial surface. These satellites exhibit brief orbital periods, typically between 90 and 120 minutes, enabling them to complete up to 16 orbits of Earth within a single day. LEO satellites are particularly advantageous for a diverse array of applications, including remote sensing, high-resolution Earth observation, and scientific inquiry, attributable to their capability for rapid data acquisition and transmission. All classifications of LEO satellites possess the ability to adjust their orientation in relation to the plane of the Earth's surface. Due to the relatively proximate distance of LEO satellites to Earth,

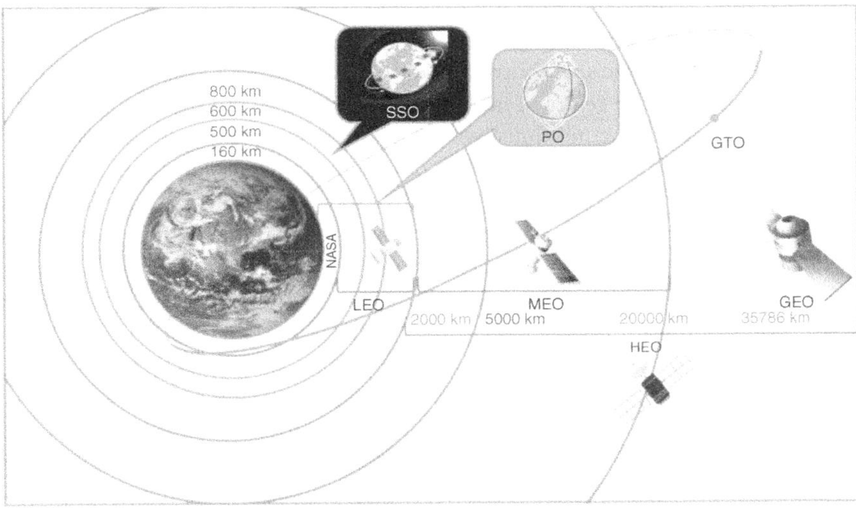

Figure 1.2 Classification of satellite orbits.

Table 1.1 Satellite altitude and operational period.

Satellite type	LEO	SSO	MEO	GTO	GEO
Altitude (km)	500–2,000	600–800	2,000–20,000	up to 35,786	35,786
Period (minutes)	up to 129	96–100	720	630	1,436

their coverage area is comparatively smaller when juxtaposed with other satellite classifications. Typically, a collective of LEO satellites, referred to as a constellation, is deployed concurrently. Collectively, LEO satellites possess the capacity to cover extensive geographical regions simultaneously [6].

MEO satellites are situated between LEO and GEO, generally at altitudes ranging from approximately 2,000 to 20,000 km. Positioning and navigation systems, such as the global positioning system (GPS), significantly rely on MEO satellites for their operational functionality. Recently, the HTS MEO constellation has been implemented to facilitate low-latency data communication services. MEO satellites exhibit an extended orbital period, varying between 2 and 12 hours, which establishes a balance between the coverage area and the data transmission rate. MEO satellites necessitate a reduced amount of infrastructure to achieve global coverage compared to LEO satellites; however, they experience longer latency and exhibit weaker signal strength.

When viewed from terrestrial vantage points, objects situated in GEO manifest an apparent state of permanence. The orbital period of GEO aligns precisely with the rotational period of Earth—specifically, 23 hours, 56 minutes, and 4 seconds—thereby enabling ground-based antennas to maintain continuous alignment with the same spaceborne apparatus. Furthermore, GEO satellites serve critical functions in the realm of meteorology, facilitating the continuous surveillance of atmospheric conditions within specific geographic locales and the meticulous observation of the evolution of localized meteorological phenomena. Nevertheless, a significant drawback of utilizing GEO satellites for instantaneous communication arises from their considerable distance from Earth, which results in substantial latency in signal transmission.

SSO satellites operate at altitudes ranging from approximately 600 to 800 km above the Earth's surface as they navigate the polar regions from north to south; the orbital inclination and altitude of these satellites are meticulously calibrated to guarantee that they consistently pass over any specific location at precisely the same local solar time. Consequently, the illumination conditions during imaging remain uniform, rendering this category of satellite exceptionally well-suited for missions related to Earth observation and environmental monitoring. Both contemporary and historical satellite imagery obtained from SSO are particularly effective for change detection. Researchers utilize these sequences of images to comprehend the progression of meteorological phenomena, forecast cyclones, oversee and mitigate wildfires and flooding events, as well as to collect data on persistent issues such as deforestation and alterations in shorelines. Nevertheless, due to their relatively low orbital altitude, SSO satellites are capable of covering only a limited geographic area at any given moment, thus necessitating the deployment of additional equipment to accomplish continuous coverage.

The most prevalent transfer orbit utilized for satellite deployment is the GTO, which facilitates the transition of satellites from a transitional orbit to GEO. When a launch vehicle propels a spacecraft into the cosmos, the spacecraft does not invariably proceed directly to its ultimate orbital position. Initially, the rocket will position the spacecraft into a transfer orbit, serving as an intermediary station on its trajectory toward the final destination. Subsequently, the satellite's propulsion system is activated to enable the satellite to attain its designated orbital position and modify its orbital inclination. This expedited approach permits the satellite to attain a geostationary orbit while expending minimal resources.

Moreover, there exist a variety of specialized satellites in orbit, which are delineated as follows.

The GEO satellite is situated within a highly elliptical orbit (HEO) characterized by significant orbital eccentricity, resulting in a pronounced disparity in distance between the segment of the orbit nearest to Earth and that which is more distally located. The operational altitude of an HEO satellite generally spans from approximately 2,000 to several hundred thousand kilometers. The inclination of HEO satellites is predominantly substantial, catering to specific mission exigencies. Due to their capacity to encompass the polar regions of Earth and higher latitudinal zones, HEO satellites are frequently utilized for military reconnaissance and surveillance purposes, particularly in operations that complement other orbital frameworks.

Polar orbit (PO) satellites are defined as satellites that traverse the atmospheric region above the polar extremities of Earth. Satellites positioned within such orbits intersect the Earth's northern and southern poles at a specific angular inclination. With each passage over the terrestrial surface, the satellite's trajectory establishes a particular angle relative to the Earth's rotational vector, thereby enabling a progressive coverage of diverse geographical areas. Owing to the Earth's axial rotation, the satellite experiences a minor deviation during each successive transit over the identical location, ultimately facilitating comprehensive surveillance of the entire polar region. Typically, PO satellites operate within an altitude spectrum ranging from approximately 160 to 2,000 km, exhibiting an orbital inclination approaching 90°. These satellites function at relatively elevated velocities, generally between 7.6 and 8 km per second, with an orbital period of approximately 90–100 minutes to complete one revolution around Earth. The operational characteristics of PO satellites render them particularly advantageous for military reconnaissance, surveillance, and espionage activities, attributed to their capacity to survey all global territories. They are generally equipped with advanced high-resolution imaging capabilities and can maintain prolonged observational periods, exemplified by the KH-11 series of reconnaissance satellites deployed by the United States. Furthermore, PO satellites hold substantial significance for scientific inquiry, particularly in the realm of long-term observations pertaining to geology, climate, and ecology on a planetary scale [7].

Table 1.2 Configuration parameters of the VLEO satellite.

Parameters	VLEO satellite	LEO satellite
Altitude (km)	1,500–450	500–2,000
End-to-end delay (ms)	<5	20–40
Satellite lifetime (year)	1–3	5–10
Transmission costs	< 100 kg, 500,000–1,000,000	500–1,000 kg, 500,000, or more
Atmospheric drag effects	Orbital decay of 0.5–1 km per day	None

Very low Earth orbit (VLEO) satellites utilize high-frequency bands (Ka/V-band, 40–75 GHz) to attain terabit-level data throughput; however, this incurs an additional atmospheric attenuation of 15–20 dB in comparison to LEO, as illustrated in Table 1.2. In order to alleviate this issue, adaptive coding and modulation (ACM) algorithms are employed to dynamically modify transmission parameters, a strategy exemplified within the National Aeronautics and Space Administration's (NASA's) Space Communications and Navigation program. Multi-beam phased array (PA) antennas, such as those integrated into Airbus's VLEO-COM prototype, facilitate precise beamforming (BF) with a pointing accuracy of less than 0.1, thereby compensating for signal degradation while achieving a capacity of 1 Tbps per satellite [8].

Ultra-low latency represents a crucial advantage, characterized by round-trip delays falling below 5 ms at an altitude of 300 km—imperative for financial trading activities and autonomous systems. In order to mitigate spectrum congestion, cognitive radio systems, propelled by artificial intelligence (AI) and exemplified by DARPA's prototypes from the Spectrum Collaboration Challenge, facilitate real-time interference mitigation within shared frequency bands [9]. Dual-circular polarization antennas, utilized in China's Galaxy space VLEO test satellites, enhance spectral efficiency threefold through the polarization multiplexing technique. The implementation of onboard AI autonomy is poised to transform operational methodologies. Machine learning (ML) algorithms, such as the European Space Agency's (ESA's) CyberSeed, empower autonomous collision avoidance mechanisms and network self-healing capabilities, allowing for the reconfiguration of routing paths within a mere 10 seconds following the occurrence of satellite failure.

Quantum key distribution (QKD) can support a speed of 10 kbps over a distance of 1,000 km. Moreover, as reported by *Nature* on 19 March 2025, researchers have made a groundbreaking advancement in quantum communication by achieving QKD over an impressive distance of 13,000 km, from China to South Africa, using the lightweight, cost-effective microsatellite, Jinan-1 [10]. This satellite transmitted laser pulses in quantum states, generating a secure quantum key that ensures confidentiality even in the presence of future quantum computing threats. Compared to its predecessor, Micius, Jinan-1 is significantly smaller, more affordable, and more efficient, with a greatly reduced ground-station receiver. This breakthrough represents a major step toward establishing a global QKD network, making transmissions virtually impervious to quantum computing threats. Although Jinan-1 does not yet support advanced features such as entangled photon generation, future miniaturization developments could enable these capabilities, further advancing the creation of a global quantum Internet. In-orbit manufacturing technologies, exemplified by Airbus's CIMON 3D-printing initiative, are designed to fabricate satellite structures directly in the extraterrestrial environment, circumventing launch mass limitations and extending operational lifespans to a duration of 5 years.

Numerous pioneering projects demonstrate the commercial feasibility of VLEO:

- SpaceX's VLEO Starlink Layer: Functioning at an altitude of 340 km and utilizing laser inter-satellite links (ISL), it achieves a latency of less than 4 ms, thereby facilitating high-frequency trading as well as autonomous vehicle networks.
- Lockheed Martin's Pony Express: A constellation positioned at 450 km that incorporates QKD-enabled military communications, which exhibit a high degree of resistance to eavesdropping attempts.
- NorthStar's Skylark: A network composed of 60 satellites operating at 320 km, delivering 20 Mbps broadband services to Arctic shipping routes at a cost of 50 dollars/GB, utilizing ice-resistant 28 GHz antenna technology.

1.2.2 Developments and Future Evolution of Satellite Communication

Satellite communication is currently experiencing a significant transformation, propelled by advancements in technology, evolving market requirements, and the imperative for worldwide connectivity. Prospective advancements will prioritize enhanced throughput, diminished latency, comprehensive global coverage, sustainability, and seamless integration with terrestrial communication networks. The swift implementation of LEO mega-constellations, as illustrated by initiatives such as SpaceX's Starlink, OneWeb, and China's GuoWang, signifies a fundamental shift in the domain of global satellite communication. These constellations employ cutting-edge technologies—including PA antennas for agile beam steering, optical ISLs (OISLs) for expedited data transmission, and non-terrestrial network (NTN) integration compliant with the 3rd generation partnership project (3GPP)—to provide unparalleled low-latency (sub-50 ms) and high-throughput connectivity. SpaceX's Starlink stands as the preeminent satellite Internet provider globally, having launched over 6,000 satellites by July 2024, with aspirations to increase this number to 42,000 in its Generation 2 constellation. Its services encompass the majority of the globe, although polar regions remain under regulatory constraints. OneWeb is set to finalize its constellation deployment of 648 satellites in 2024 and has merged with Eutelsat in 2023, thereby integrating GEO and LEO capabilities. This initiative is predominantly targeted toward enterprise and government sectors (for instance, maritime, aerospace, and defense), with collaborators that include Airbus, Hughes, and Airtel. The technology employed is Ku-band, and the terminals are specifically engineered for high-mobility applications. Amazon's Project Kuiper is poised to launch 3,236 satellites by 2029 and has already initiated its first test satellites in 2023. Its innovations encompass cost-effective terminals (with an anticipated 400–500 consumer terminals utilizing PA antennas). Telesat envisions the deployment of 298 satellites, primarily focusing on the enterprise and government sectors, placing particular emphasis on low latency (30–50 ms), which is especially advantageous for financial transactions and telecommunications backhaul. The technology utilized incorporates advanced PA antennas and optical interplanetary links. China's GuoWang Constellation intends to deploy over 13,000 satellites to compete with Starlink. This initiative is supported by state-owned enterprises, such as the China Satellite Network Group, and aims to facilitate domestic and Belt and Road Initiative connectivity, with a pronounced focus on military-civil fusion applications [11].

The HTS systems utilizing elevated frequency bands such as Q/V (40–50 GHz) and Ka/Ku facilitate terabit-level data throughput [12], thereby substantially augmenting data capacity. Software-defined satellites equipped with reconfigurable payloads proficiently allocate bandwidth and power in real-time, thereby optimizing operational performance in accordance with demand fluctuations. OISLs further mitigate latency by facilitating laser-based communication between satellites, thus circumventing the necessity for ground station involvement. These technological advancements yield significant advantages for both enterprise and governmental sectors by delivering ultra-secure, high-capacity communication links for industries including mining, energy, and defense, while concurrently addressing the digital divide in underserved and rural locales.

The amalgamation of satellites with 5th generation mobile networks (5G) and 6th generation mobile networks (6G) alongside terrestrial networks is facilitating a seamless integration of connectivity. Through 3GPP NTN, satellite technology is poised to become an intrinsic element of 6G, thereby enabling uninterrupted roaming capabilities between cellular and satellite infrastructures. The implementation of edge computing on satellites will permit on-orbit data processing, thereby diminishing latency for Internet of Things (IoT) applications, which encompass autonomous vehicles and disaster management operations. Hybrid frameworks that integrate GEO, MEO, and LEO satellites with terrestrial 5G small cells will establish a cohesive "network of networks," thereby augmenting global connectivity [13]. Practical applications such as emergency communication systems for disaster-stricken areas and smart agricultural initiatives utilizing IoT sensor connectivity in remote farming locales underscore the transformative potential of these integrated systems.

Emerging technologies in the realm of quantum communication are poised to fundamentally transform secure data transmission methodologies. QKD, as exemplified by China's Micius satellite, facilitates ultra-secure encryption through the application of quantum mechanical principles [14], rendering it virtually infeasible to intercept or compromise communication without being detected. Concurrently, experimental endeavors in satellite-based quantum entanglement are laying the groundwork for the establishment of global quantum networks, which have the potential to substantially improve both the security and efficacy of communication systems. These advancements hold the promise of delivering unbreakable encryption, thereby guaranteeing the utmost levels of security for sensitive data and communications on a global scale.

Innovations in AI are revolutionizing satellite operations, thereby enhancing both operational efficiency and autonomy [15]. Resource allocation driven by AI employs ML algorithms to optimize essential satellite functions, including bandwidth, power, and routing, in real time, thereby ensuring optimal performance and dynamic adaptability to varying network conditions. Autonomous collision avoidance systems, which are fueled by AI, enable satellites to anticipate potential collisions with space debris and execute real-time adjustments, effectively mitigating the risk of catastrophic incidents without necessitating human intervention. Furthermore, predictive maintenance algorithms persistently monitor the operational health of satellites, utilizing data to anticipate potential failures and initiate preemptive measures, consequently minimizing downtime and prolonging the operational lifespan of the satellite fleet. These advancements in AI are crucial for attaining more efficient, secure, and sustainable satellite operations in the foreseeable future.

The commercial satellite industry is currently experiencing a significant transformation influenced by advancements in technology and the dynamics of the market. The increasing prevalence of small satellites, encompassing CubeSats and nanosatellites, has facilitated broader access to space, allowing for the economically viable deployment of extensive mega-constellations such as SpaceX's Starlink and OneWeb [16]. These satellite constellations provide low-latency broadband services to remote and underserved areas, thereby diminishing the digital divide while simultaneously lowering per-unit expenses through the principles of economies of scale. Concurrently, advancements in additive manufacturing (3D printing) and the design of modular satellites are optimizing production processes, leading to cost reductions of up to 40%, and fostering the rapid prototyping of reusable components—elements that are essential for the expansion of next-generation networks.

As the phenomenon of orbital congestion intensifies, the mitigation of space debris has emerged as a paramount strategic concern. Current assessments suggest that there are in excess of 130 million debris fragments within LEO, thereby necessitating the implementation of sophisticated deorbiting mechanisms such as drag sails, electrodynamic tethers, and active removal systems, exemplified by the ESA's ClearSpace-1 initiative. The objective of these technologies is to ensure adherence to the 25-year deorbiting guideline and to mitigate the potential for Kessler Syndrome—a catastrophic scenario characterized by a chain reaction of collisions that could jeopardize the usability of essential orbital paths. Regulatory frameworks are undergoing evolution to enforce end-of-life disposal protocols, while AI-driven collision avoidance systems are being assimilated into satellite operations to augment real-time decision-making capabilities.

The commercial exploration of space is experiencing a marked intensification, characterized by competition among nation-states and corporate entities to capitalize on lunar resource extraction and establish advantageous positions within LEO. Private sector organizations are at the forefront of developing in-situ resource utilization technologies aimed at lunar ice extraction, whereas governmental agencies are utilizing satellite constellations for dual-purpose applications, encompassing Earth observation alongside secure military communications. This competitive landscape reflects dynamics reminiscent of the Cold War era; however, it is now significantly propelled by private financial investments, which have surpassed $10 billion annually [17]. As regulatory agencies contend with the complexities of spectrum allocation and the management of orbital slots, the industry's emphasis on sustainability, cost-effectiveness, and technological adaptability will be critical in determining its ability to reconcile growth with the safeguarding of the extraterrestrial environment.

1.2.3 Basic Patterns of Satellite Communications

With the persistent escalation of demand and advancements in technology, satellite communication has transcended its conventional confines of terrestrial space communication and has progressively evolved into a multifaceted communication system encompassing interstellar communication, collaborative multi-satellite communication, inter-orbit satellite communication, and deep-space communication across a diverse array of modalities.

Satellite-ground communication represents the quintessential and foundational modality of satellite communication, encompassing the bidirectional transfer of data between satellites and terrestrial ground stations. The primary operation of satellite-ground communication entails the transmission of signals from the ground station to the satellite (uplink) and the subsequent return of these signals to the terrestrial domain (downlink) [18]. The hardware and apparatus associated with satellite-to-ground communication systems have been subject to continuous enhancements aimed at augmenting efficiency, reliability, and coverage capabilities. The satellite payload constitutes the principal component of the satellite tasked with the reception, processing, and retransmission of signals, wherein the radio frequency (RF) link serves as the medium for signal transmission and reception. Concurrently, the modem facilitates the modulation and demodulation of signals, thereby accommodating diverse communication modes and data rates. Through the application of modulation methodologies such as quadrature phase shift keying (QPSK) and 16-quadrature amplitude modulation (16QAM), in conjunction with coding strategies like low-density parity-check (LDPC) code and turbo code, the modem enhances the data transmission rate and resilience against interference. It is the satellite's antenna that is charged with the reception of signals from terrestrial ground stations in the orbital environment, transmitting these signals to the satellite payload and relaying the processed information back to the ground. Satellite antennas typically employ PA antennas or beam steering antennas, which possess the capability for precise beam alignment and multi-beam operational functionality. Furthermore, the ground station serves as both the initiation and conclusion point for star-earth communication, equipped with a high-gain parabolic antenna [19]. The ground station's antenna can be accurately oriented toward the satellite, thereby ensuring optimal signal transmission quality. The diameter of these antennas is typically substantial, often spanning several meters or even tens of meters.

Inter-satellite communication pertains to the direct transmission of data between satellites via wireless communication channels. In contrast to conventional satellite-to-ground communications, inter-satellite communications eliminate the necessity for terrestrial ground stations by facilitating the forwarding and transmission of signals through direct interlinks among multiple satellites. Data is exchanged among satellites through ISLs, typically employing microwave or laser communication methods. The function of inter-satellite communication is to augment the flexibility, scalability, and reliability of satellite networks, particularly within LEO and MEO satellite constellations. A quintessential application scenario for inter-satellite communication can be observed in LEO constellations, such as SpaceX's Starlink and OneWeb, among others. By utilizing inter-satellite communication, LEO satellites can directly exchange information with one another without reliance on ground relay, thereby establishing a vast, interconnected satellite network.

In conventional single-satellite communications, the integrity of the signal is predominantly influenced by the spatial separation between the satellite and the terrestrial station, the design specifications of the antenna, ambient environmental noise, and the phenomenon of signal attenuation. The implementation of multi-satellite collaborative communication significantly enhances both the quality of the signal and the capacity for coverage across multiple dimensions through the synergistic efforts of multiple satellites. The interference arising from various satellites presents a considerable challenge within multi-satellite collaborative networks. Through the utilization of adaptive signal processing techniques and spectrum multiplexing technology, multiple satellites are proficiently able to circumvent interference amongst themselves, thereby facilitating the dynamic allocation of spectral resources. The malfunction of a solitary satellite has the potential to precipitate a cessation of communication. Conversely, in the multi-satellite cooperative communication framework, the presence of multiple satellites functions as a redundancy mechanism for one another, allowing for swift transition and management by alternative satellites in the event of a satellite's failure or operational dysfunction. Even in scenarios where the communication link of a single satellite is subjected to interference or damage, the remaining satellites can sustain the continuity of communication through rapid link switching, thus mitigating the risk associated with a singular point of failure within the system. An ensemble of satellites outfitted with multi-beam antennas collaborates to augment signal coverage by leveraging the angular disparities present in the spatial domain. The concurrent coverage of distinct regions by multiple beams emanating from each satellite not only enhances signal strength but also effectively diminishes the adverse effects of environmental factors, such as atmospheric attenuation and interference.

Inter-orbit satellite communications refer to the interactions among satellites positioned in distinct orbital paths, particularly the exchanges occurring between LEO satellites and GEO satellites. Given that LEO satellites are characterized by reduced signal transmission latency and augmented data throughput capabilities, while GEO satellites offer extensive coverage areas, the integration of inter-orbit satellite communications can effectively amalgamate the respective advantages of both categories to facilitate efficient global communications. GEO satellites are capable of functioning as pivotal relay stations to support LEO satellites in extending their coverage across broader global regions [20].

Deep-space communications encompass the transmission of signals throughout the solar system and into interstellar space, specifically concerning celestial bodies or spacecraft and their corresponding ground control centers. The principal attributes of this communication framework include exceedingly extensive transmission ranges, significant signal degradation, as well as the complexities posed by exceptionally high latency and diminished signal strength. The technology utilized for deep-space communication is extensively applied in missions aimed at planetary exploration (e.g., NASA's Mars rovers), remote sensing in deep space, and communications associated with space station operations.

1.2.4 Development of Satellite Bands

The frequency bands designated for satellite communication are typically categorized into various segments based on the frequencies of electromagnetic radiation, with each segment exhibiting distinct characteristics and applications as delineated in Table 1.3. In accordance with the definition posited by the International Telecommunication Union (ITU), the frequency bands employed by satellite communication systems (SCSs) principally encompass the L-band through the Ka-band and extend to higher frequency bands. These frequency bands are optimized for a variety of application contexts, including but not limited to communications between terrestrial ground stations and satellites, inter-satellite connectivity, and backhaul links originating from satellites to terrestrial infrastructures [21]. The subsequent sections delineate the frequency bands that are prevalently utilized in satellite communication, along with their associated nomenclature.

In recent years, the evolution of technological capabilities has led to an increased scholarly focus on the terahertz (THz) band, particularly within the domains of satellite communications and air-to-air communications, wherein it facilitates electromagnetic wave bands characterized by frequencies ranging from 0.1 to 10 THz. A principal advantage associated with the THz band lies in its exceptionally large bandwidth, which is adept at accommodating extremely high data transmission rates; however, it concurrently contends with the impediments posed by atmospheric absorption and attenuation.

Laser communication systems, employing visible and near-infrared light wavelengths, represent a significant advancement for prospective satellite communication architectures. Through the utilization of laser links, data transmission rates

Table 1.3 Applications of different frequency bands of satellites.

Frequency band	Frequency range	Applications
L-band	1.53–1.72 GHz	Low-speed data, navigation
S-band	2.0–4.0 GHz	Weather radar, aviation
C-band	4.0–8.0 GHz	Satellite TV, military
X-band	8.0–12.0 GHz	Military, government
Ku-band	12.0–18.0 GHz	Satellite Internet, TV
Ka-band	26.5–40.0 GHz	High-speed Internet, broadband
V-band	40–75 GHz	High bandwidth, point-to-point
W-band	75–110 GHz	High bandwidth, military

can reach 1 Gbps, exceed 10 Gbps, and potentially achieve Tbps. This system is comprised of a laser transmitter, a receiver, an optical targeting mechanism, and a data processing unit, which collectively facilitate effective communication within an exceptionally compact framework [22].

1.3 Applications of Satellite Communications

In contemporary satellite communication frameworks, the deployment of the integrated sensing, communication, and computation (SCC) network has emerged as a pivotal factor in enhancing the performance, efficiency, and dependability of the system. By leveraging AI and edge computing technologies, the integrated terminal dismantles the technical barriers that exist among SCC through effective data fusion and collaborative computation, thereby optimizing the efficacy of the SCS. Through its network architecture, the terminal device not only executes data sensing but also facilitates instantaneous communication and local computation, thereby diminishing the latency of data transmission and alleviating bandwidth constraints, while enhancing the stability and responsiveness of the network.

As illustrated in Figure 1.3, in the context of satellite communication scenarios, the terminal acquires data across diverse domains, including aerial, terrestrial, and maritime environments, employing an array of sensing devices. These data encompass satellite signal strength, meteorological parameters, user geographic coordinates, and equipment operational status. The pay-per-click (PPC) network engages in the expeditious analysis and processing of extensive sensory data utilizing edge computing and AI technologies. On this premise, the PPC network is proficient in eliminating superfluous data and extracting pertinent information, thereby furnishing more precise inputs for subsequent communication and computational processes. By intricately integrating perception data with information derived from the communication system, the satellite terminal is capable of executing target identification and localization with high accuracy in more intricate environments. In adverse climatic or dynamic conditions, the synergistic interplay of perception and communication endows satellite communication with the capacity to adjust communication parameters while ensuring the reliable transmission of signals. Edge computing represents a fundamental technology within the integrated SCC network, facilitating computations on the terminal side and diminishing the duration required for data transmission to distant central servers. Consequently, edge computing enhances the speed and efficiency of data processing, particularly in scenarios characterized by high real-time demands, such as autonomous driving and the industrial Internet. In the realm of autonomous driving, the satellite communication terminal employs edge computing to analyze surrounding environmental data, including road conditions, traffic signage,

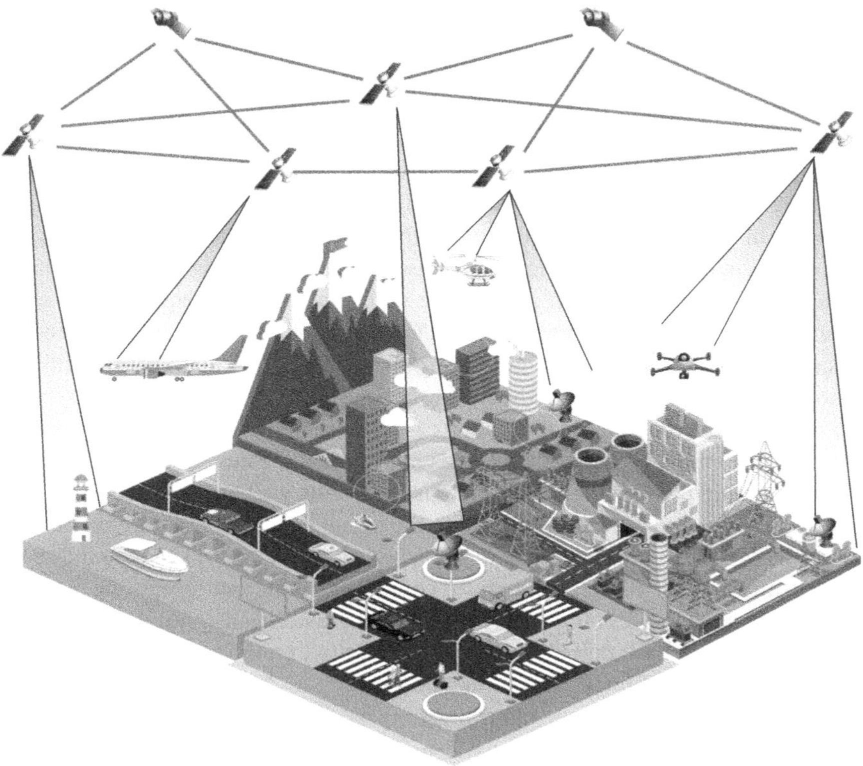

Figure 1.3 Satellite communication application scenarios.

and obstacle information, while concurrently communicating with the satellite in real-time regarding navigation and control. Furthermore, through the implementation of edge computing, the driving system is enabled to make rapid decisions, thereby safeguarding the safety and efficacy of the self-driving vehicle.

The utilization of AI technology within satellite communications is primarily manifested in the enhancement of communication protocols and the determination of signal transmission pathways. AI possesses the capability for pattern recognition of acquired data via deep learning (DL) algorithms, thereby enabling the forecasting of network traffic, the modification of channel resources, and even the preemptive identification of potential interference factors. By integrating sensing and computational abilities, AI markedly augments the adaptability and robustness of satellite communication networks. In the context of channel estimation within SCSs, AI can dynamically modulate spectrum resources and transmission power by assimilating historical communication data and environmental information to guarantee the quality and stability of communication. Furthermore, AI adeptly optimizes the communication link by autonomously adjusting the channel in response to real-time environmental fluctuations, thereby enhancing the overall performance of the entire communication link.

It is essential to acknowledge that the integrated SCC network necessitates not only the amalgamation of terminal and edge computing paradigms but also engages in a profound interaction with the central cloud server. By leveraging resource synergy, the central cloud server is capable of executing dynamic resource scheduling and allocation in accordance with the real-time requirements emanating from terminals and edge nodes. The central cloud server employs mechanisms such as channel estimation and resource allocation to adaptively modify the resource allocation strategy of the network, contingent upon the demand scenarios presented by the satellite terminals (e.g., requirements for bandwidth, computational power, etc.). Consequently, this approach mitigates the risk of resource wastage within the satellite network; for instance, during periods of elevated load, the system can dynamically re-calibrate the signal coverage area or enhance the transmission path to minimize communication latency and alleviate bandwidth constraints. The merits of the integrated communications and computing network of satellites are manifested not only in their data fusion and computational capabilities but also in their potential to foster the iterative enhancement of system performance via continuous self-optimization processes. Satellite communications facilitate a plethora of applications, encompassing cognitive radio for dynamic spectrum sharing, stratospheric platforms aimed at IoT integration, smart city infrastructures, and space-based energy transmission systems.

1.4 A Brief Description of Security Issues in Satellite Communications

1.4.1 Security Risks of Satellite Communications

The satellite communication infrastructure encounters a multitude of security vulnerabilities, which include those emanating from weaknesses inherent in satellite communication protocols as well as the potential misuse of emerging technologies. In addition to the security challenges analogous to those encountered by conventional terrestrial mobile networks, the distinctive security threats illustrated in Figure 1.4 relevant to SCSs predominantly comprise interference, unauthorized eavesdropping, and assaults targeting radio links, payloads, and the satellite platforms associated with SCSs.

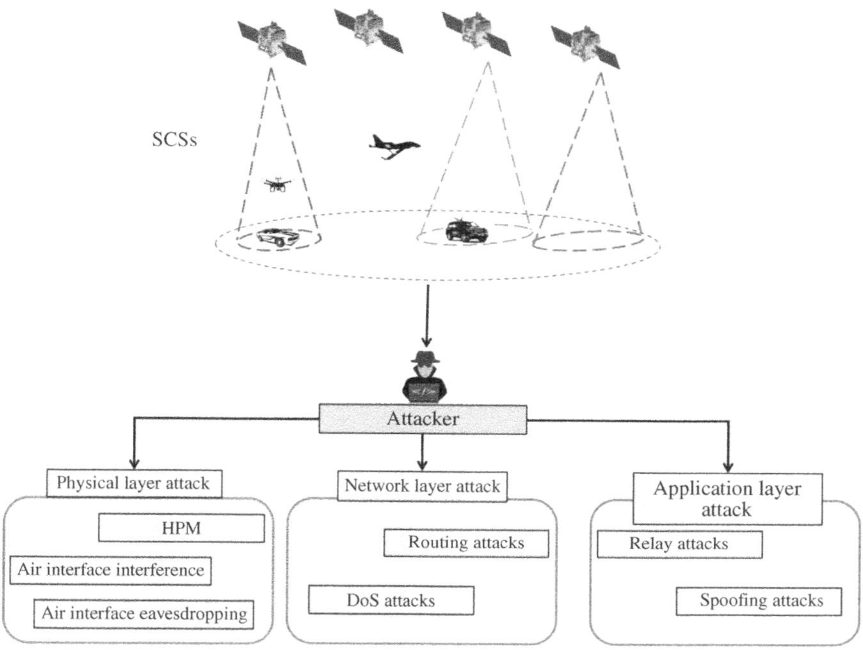

Figure 1.4 Satellite security communications model.

The vulnerability of legitimate user signals to malicious jamming is exacerbated by the inadequate wireless signals received by all devices within the satellite communication framework. The disruption of user links, feeder links, and interstellar links may precipitate the cessation of communication services within specific geographic locales. Concurrently, interference affecting measurement and control links results in the temporary incapacitation of telemetry, remote control, and various operational functions, thereby undermining the regular functioning of the satellite. The interference encountered by SCSs can be systematically classified into suppression interference and deception interference. The inherent openness and extensive coverage of satellite communications render them particularly susceptible to air-interface eavesdropping. Given the broad spectrum of downlink beams, adversaries frequently possess the capability to intercept signals from considerable distances, facilitating the eavesdropping or interception of communication content, especially in the absence of robust encryption protocols. Adversaries not only have the capacity to capture signals via antennas but may also exploit advanced signal processing methodologies to decode the transmitted communication content. In the future, satellite communications are anticipated to encounter increasingly intricate environments, wherein not only the challenges associated with physical eavesdropping but also the dynamic interaction between satellite and terminal apparatus, along with signal interference, must be effectively mitigated.

The peril associated with a high-power microwave assault lies in its capacity to obliterate satellite apparatus and generate extensive electromagnetic interference phenomena. Through the transmission of high-power pulses aimed at the designated satellite, the assailant effectively annihilates the electronic components within the satellite's reception system and may even incapacitate its measurement and control infrastructure. In contrast to conventional cyberattacks, this form of assault results in the physical destruction of the satellite system, thereby inducing irreversible harm. It is imperative to acknowledge that, given the escalating complexity of satellite functionalities, the implementation of protective strategies against high-power microwave assaults must increasingly emphasize electronic anti-jamming technologies, redundant satellite architecture, and comprehensive electromagnetic compatibility.

The pathways for data transmission within satellite networks exhibit considerable diversity and complexity. Malicious actors fabricate spurious routes within spatial links by either falsifying or manipulating routing data. The repercussions of routing attacks in satellite networks are significantly more pronounced than in conventional networks, attributable to the prolonged propagation times and limited fault tolerance inherent in satellite links, which exacerbate routing inaccuracies. Nevertheless, in the context of multi-constellation or multi-link satellite networks, adversaries are capable of not only manipulating the data flow through the falsification of routing information but also executing aggressive distributed denial-of-service (DDoS) assaults. To mitigate such threats effectively, SCSs must implement robust routing authentication protocols and integrate dynamic monitoring alongside intelligent response mechanisms. Within the dynamic architecture of satellite Internet, the authentication challenges between nodes represent a notable vulnerability in terms of security. Intruders infiltrate the satellite network by masquerading as legitimate nodes, thereby causing disruptions in network data transmission or complete operational paralysis. In contrast to conventional computer network intrusions, the highly distributed nature and low latency prerequisites of satellite communications amplify the potential severity of these attacks. By impersonating an earth station or satellite node and engaging in network activities, the integrity of communication is compromised, and data flow is redirected toward illicit pathways. Consequently, in addition to authentication measures, there is a pressing need to enhance the dynamic security management of nodes through the integration of multi-factor authentication (MFA) and behavioral monitoring methodologies.

The satellite communication platform is characterized by constrained resources, and adversaries emulate legitimate terminals to initiate a substantial volume of random access requests, thereby monopolizing the channel resources at the physical layer. Nevertheless, this phenomenon represents merely the preliminary aspect of "denial of service (DoS)." As adversaries persist in augmenting their jamming capabilities, the repercussions on satellite communications are becoming progressively pervasive. Furthermore, attackers possess the ability to compel satellites to function within the nonlinear domain by deploying high-power interference, which consequently engenders a power-skimming effect. This can inhibit normal signals and potentially incapacitate the entire satellite communication network. In contrast to conventional network environments, the extensive link spans, stringent low-latency requirements, and issues associated with single points of failure present in satellite systems exacerbate the lethality of these attacks on satellite communications. The peril associated with replay attacks is frequently undervalued, particularly in scenarios where satellite platforms are constrained by limited resources. Numerous satellite measurement and control links are inadequately equipped to address the threat posed by replay attacks. By intercepting and archiving commands transmitted between satellites and ground stations, adversaries can repetitively issue these commands. Indeed, not only is the command level vulnerable, but also orbit control, antenna positioning, and even the allocation of satellite resources may be jeopardized by replay attacks. In instances where commands are not safeguarded by robust authentication or time-stamping mechanisms, satellites risk executing redundant operations. This threat is especially acute in military or strategic communication contexts and necessitates the implementation of more rigorous security protocols, along with real-time monitoring and defensive strategies to alleviate its potential impact.

1.4.2 Security Protocols and Mechanisms of Satellite Communication

Owing to the complexities inherent in historical evolution and the diverse requirements of various industries, it proves to be a formidable task to pinpoint a singular SCS that completely conforms to universally recognized standards. Nevertheless, the majority of modern SCSs are constructed upon international standards established by entities such as the Consultative Committee for Space Data Systems (CCSDS), the European Telecommunications Standards Institute (ETSI), and 3GPP. These standards present an array of alternatives that are specifically designed to accommodate distinct user requirements and operational contexts.

The Space Communication Protocol Specifications (SCPS), introduced by the CCSDS during the 1990s, primarily concentrated on the communication protocols pertaining to the link and network layers. Subsequently, the TCP/IP was incorporated to facilitate the interconnection of various layers within the SCPS architecture, thereby enabling a more holistic approach to communication. The principal design objective of SCPS was to address the resource-constrained demands inherent in satellite communication by minimizing communication overhead while optimizing bit rates. The security protocol associated with SCPS is dedicated to ensuring the security of communications between the network and transport layers, employing data encapsulation methodologies to achieve authentication and access control. The provision of services such as data integrity verification, mechanisms for confidentiality, and identity authentication is contingent upon the specific requirements of the users. Notwithstanding its provision of minimal communication overhead and enhanced transmission efficiency, the defensive measures of the SCPS protocol against replay attacks are still markedly inadequate.

Digital Video Broadcasting-Return Channel via Satellite (DVB-RCS), which was established by the ETSI in 2000, represented the inaugural standard for interactive satellite communication applications. In 2012, DVB-RCS2 was introduced, incorporating substantial enhancements in the foundational protocols, higher-layer functionalities, and security frameworks. Since its inception, DVB-RCS2 has achieved widespread implementation in Ku-band and Ka-band broadband SCSs.

In the 5G communication framework, 3GPP delineates an extensive security architecture that caters to the requirements of satellite communications, encompassing user-plane confidentiality, control-plane integrity, anti-replay protocols, access authentication, and key negotiation processes. These security provisions collectively ensure the confidentiality and integrity of data transmissions while mitigating potential threats such as replay attacks and identity impersonation.

1.5 Organization of This Book

The framework of this book is carefully orchestrated to facilitate a comprehensive examination of satellite communication frameworks, emphasizing their security, architectural design, and prospective innovations. The composition is segmented into five sections, each addressing unique yet interrelated subjects essential for grasping the field.

Part 1: Fundamentals commences with an examination of the historical context surrounding satellite communications, presenting a succinct yet thorough analysis of their progression, utilization, and the emerging security dilemmas associated with them. Chapter 1 familiarizes the reader with the historical framework, advancements, and extensive applications of satellite communications while simultaneously offering an initial comprehension of the intrinsic security issues pertinent to these systems. Subsequently, Chapter 2 shifts focus to the architectural framework of satellite communications, delineating the functions of space, terrestrial, and user segments, in addition to elucidating the notion of terrestrial coverage areas.

Part 2: Security issues and requirements provides an in-depth examination of the particular security threats that confront SCSs. Chapter 3 delineates the diverse categories of security risks, encompassing both active and passive attacks, and scrutinizes the vulnerabilities inherent within satellite systems. It further confronts reliability challenges, including interference and single-event upsets (SEUs). Expanding upon this foundation, Chapter 4 analyzes the critical security requirements pertinent to satellite communication, emphasizing essential performance metrics such as confidentiality, integrity, accountability, and system reliability with respect to latency, capacity, and throughput.

Part 3: Security enhancement solutions concentrates on prospective methodologies aimed at augmenting the security framework of satellite communications. In Chapter 5, active security enhancement methodologies are examined, encompassing pioneering technologies such as quantum communication, blockchain systems, THz communication, spaceborne laser systems, and AI applications. Conversely, Chapter 6 emphasizes passive security methodologies, including sophisticated antenna designs, reconfigurable intelligent surfaces (RISs), and techniques for interference coordination. The collaboration of satellites and the application of AI are also evaluated to bolster reliability and security measures. Chapter 7 redirects focus toward enhancements related to security-driven reliability, addressing critical elements such as outage mitigation, bit error rate (BER) management, and forward error correction (FEC), with AI playing a pivotal role in system optimization.

Part 4: Design criteria for security SCSs offers a comprehensive technical analysis regarding the development of secure and dependable SCSs. Chapter 8 delineates essential design parameters, including the selection of orbital positions and the management of frequency allocations. In Chapter 9, the functionalities of spaceborne payloads, encompassing on-board processing (OBP) and transparent forwarding, are meticulously scrutinized. Chapter 10 investigates the amalgamation of security and reliability within system design, underscoring the necessity of concurrently addressing both dimensions throughout the design process.

Part 5: Future directions examines the prospective advancements in satellite communication technologies alongside their dynamic security requirements. Chapter 11 evaluates next-generation secure transmission methodologies, including integrated sensing and communication (ISAC) assisted and computer vision (CV) assisted approaches, which bear significant implications for the forthcoming communication systems. Chapter 12 investigates the architecture and security frameworks of mega-constellations, addressing both single-layer and multilayer constellations and their secure inter-orbit communication protocols. Lastly, Chapter 13 scrutinizes the critical role of physical layer security (PLS), encompassing key generation, authentication, and encryption methodologies, all of which are essential for the protection of satellite communications within an increasingly intricate environment.

Through a methodical examination of the fundamental elements inherent in satellite communication, including the associated security challenges, strategies for enhancement, and design considerations, this scholarly work provides an extensive resource that integrates theoretical principles with practical implementations while concurrently delineating the prospective trajectory for secure and dependable SCSs.

References

1 T Teichmann. Some elementary considerations of satellite-earth communication systems. *IRE Trans. Space Electron. Telem.*, SET-7(2):51–54, Jun. 1961.

2 Walid Saad, Mehdi Bennis, and Mingzhe Chen. A vision of 6G wireless systems: Applications, trends, technologies, and open research problems. *IEEE Netw.*, 34(3):134–142, May 2020.

3 Oltjon Kodheli, Eva Lagunas, Nicola Maturo, et al. Satellite communications in the new space era: A survey and future challenges. *IEEE Commun. Surv. Tutor.*, 23(1):70–109, Oct. 2021.

4 Mauro De Sanctis, Ernestina Cianca, Giuseppe Araniti, Igor Bisio, and Ramjee Prasad. Satellite communications supporting internet of remote things. *IEEE Internet Things J.*, 3(1):113–123, Feb. 2016.

5 Raymond J Leopold. Motorola's IRIDIUM system: Low-earth orbit global cellular communications network. In *IEEE Antennas and Propagation Society International Symposium 1992 Digest*, volume 3, pages 1270, Chicago, IL, USA, 1992.

6 Alessandro Vanelli-Coralli, Alessandro Guidotti, Tommaso Foggi, Giulio Colavolpe, and Guido Montorsi. 5G and beyond 5G non-terrestrial networks: Trends and research challenges. In *2020 IEEE 3rd 5G World Forum (5GWF)*, pages 163–169. IEEE, 2020.

7 Athanasios D Panagopoulos, Pantelis-Daniel M Arapoglou, and Panayotis G Cottis. Satellite communications at KU, KA, and V bands: Propagation impairments and mitigation techniques. *IEEE Commun. Surv. Tutor.*, 6(3):2–14, Dec. 2004.

8 Xuewen Luo, Hsiao-Hwa Chen, and Qing Guo. LEO/VLEO satellite communications in 6G and beyond networks–technologies, applications, and challenges. *IEEE Netw.*, 38(5):273–285, Sep. 2024.

9 Zhe Niu, Hui Yang, Qiuyan Yao, et al. Reliable low-latency routing for VLEO satellite optical network: A multiagent reinforcement learning approach. *IEEE Internet Things J.*, 12(3):2309–2321, Feb. 2025.

10 Li Yang, Cai Wen-Qi, Ren Ji-Gang, et al. Microsatellite-based real-time quantum key distribution. *Nature*, 640:47–54, Mar. 2025.

11 Xingfa Gu and Xudong Tong. Overview of China earth observation satellite programs [space agencies]. *IEEE Geosci. Remote Sens. Mag.*, 3(3):113–129, Sep. 2015.

12 Ramy Samy, Hong-Chuan Yang, Tamer Rakia, and Mohamed-Slim Alouini. Space-air-ground FSO networks for high-throughput satellite communications. *IEEE Commun. Mag.*, 61(3):82–87, Mar. 2023.

13 John Lodge. Mobile satellite communications systems: Toward global personal communications. *IEEE Commun. Mag.*, 29(11):24–30, Nov. 1991.

14 Andrea Conti, Robert Malaney, and Moe Z Win. Satellite-terrestrial quantum networks and the global quantum internet. *IEEE Commun. Mag.*, 62(10):34–39, Oct. 2024.

15 Shadab Mahboob and Lingjia Liu. Revolutionizing future connectivity: A contemporary survey on AI-empowered satellite-based non-terrestrial networks in 6G. *IEEE Commun. Surv. Tutor.*, 26(2):1279–1321, Jan. 2024.

16 Gianluca Furano, Gabriele Meoni, Aubrey Dunne, et al. Towards the use of artificial intelligence on the edge in space systems: Challenges and opportunities. *IEEE Aerosp. Electron. Syst. Mag.*, 35(12):44–56, Dec. 2020.

17 Gary K Noreen. Regulatory flexibility in commercial satellite communications. *IEEE Commun. Mag.*, 29(2):58–60, Feb. 1991.

18 Chenbo Hu, Hongjuan Yang, Zhiquan Zhou, et al. Outage analysis of RSMA enabled integrated satellite-terrestrial networks. *IEEE Trans. Veh. Technol.*, 74:10023–10028, Feb. 2025.

19 Yunyang Zhang, Yulun Wu, Aijun Liu, Xiaochen Xia, Ting Pan, and Xian Liu. Deep learning-based channel prediction for LEO satellite massive MIMO communication system. *IEEE Wirel. Commun. Lett.*, 10(8):1835–1839, 2021.

20 Wenqing Li, Bo Ai, Yong Niu, Zhu Han, Ning Wang, and Lei Xiong. End-to-end scheduling of space-air-ground integrated networks for high-speed railways. *IEEE Trans. Veh. Technol.*, 74:9591–9603, Feb. 2025.

21 Louis J Ippolito. Characterization of the CTS 12 and 14 GHz communications links-preliminary measurements and evaluation. In *International Conference on Communications*, volume 1, pages 12_26–12_29, 1976.

22 Rui Sen Hao, Jin Fan Zhang, Shi Chao Jin, Dun Ge Liu, Ting Jun Li, and Yu Jian Cheng. K-/Ka-band shared-aperture phased array with wide bandwidth and wide beam coverage for LEO satellite communication. *IEEE Trans. Antennas Propag.*, 71(1):672–680, 2022.

2

System Architecture of Satellite Communications

This chapter introduces the system architecture of SCSs and commences with an overview of the principal components of SCSs, including the space segment, ground segment, and user segment, as illustrated in Figure 2.1, and the main components in each segment are as follows:

- **Space Segment:** The space segment consists of various orbit satellites and inter-satellite links (ISLs), where each satellite is connected by ISLs. However, not all SCSs have ISLs; a counterexample is OneWeb.
- **Ground Segment:** The ground segment is composed of the gateway, the network control center (NCC), and ground-based stations.
- **User Segment:** The user segment includes a large number of terminals that access satellite systems via the user link.

Moreover, this chapter provides an overview of recent developments in satellite constellations and progress in NTN standardization. Then, this chapter discusses the deployment and technological advancements of constellations, such as Starlink and OneWeb. Additionally, this chapter examines the standardization of NTN as part of global communication standards, with a focus on integrating satellites with existing terrestrial networks.

After that, this chapter concludes with an examination of the intrinsic characteristics and distinctive challenges of SCSs. These include the need for frequent handover due to the rapid movement of LEO satellites, the requirement for low-latency communication, and the challenges associated with the deployment and maintenance of large-scale constellations. Systems lacking ISL face additional challenges, such as the need for substantial ground infrastructure to ensure connectivity.

The subsequent sections of this chapter will provide a comprehensive analysis of the role of SCSs in modern communication networks, with the aim of establishing a solid basis for the exploration of their technical and operational challenges.

Safeguarding Satellite Communications: Issues, Challenges, and Solutions, First Edition.
Jianping An, et al.

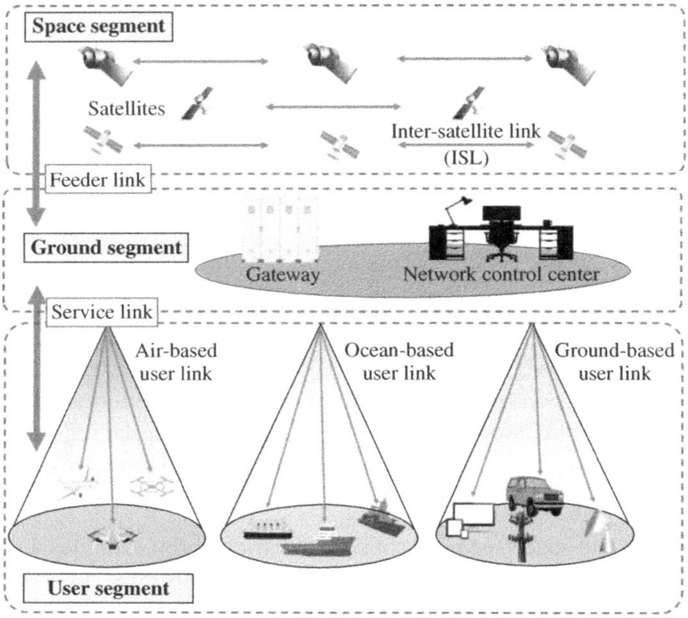

Figure 2.1 The system architecture of SCSs.

2.1 Space Part for Satellite Communication System

The space segment constitutes the fundamental component of contemporary SCS, primarily consisting of different altitude Earth orbit satellites and ISLs. This segment operates as a versatile infrastructural framework in space, enabling communication and connectivity for a diverse array of applications, such as telecommunications, remote sensing, and global networking.

This section presents a thorough investigation of the classification of satellites, their intrinsic properties and challenges, and the architectural paradigms that underpin their integration into terrestrial networks. The ensuing discourse will encompass the following key domains: Orbital altitude classification-based satellite system architecture, existing satellite projects, single- and multilayer networks (for instance, LEO–GEO integration), and the current standardization state. Furthermore, the function of diverse satellite operational modalities, such as transparent forwarding and onboard processing, will be investigated within the framework of enhancing performance and optimizing payload utilization.

2.1.1 System Architecture

Satellite networks offer pervasive coverage and uninterrupted communication services, thereby enhancing the efficiency of interactions among mobile users, while also enabling effective communication between mobile and stationary users via satellite relays. Depending on the altitude of their orbit and their operational characteristics, satellite systems can be classified into GEO, MEO, LEO, and VLEO systems. The unique properties associated with each orbital category render it particularly suitable for distinct applications.

As delineated in Table 2.1, there is a significant variation among satellite systems concerning parameters such as, but not restricted to, orbital altitude, orbital period, path loss, propagation latency, coverage area, satellite longevity, and the requisite number of satellites.

- GEO satellites, situated at the highest orbital altitude, provide the most extensive coverage while requiring minimal constellation, but they experience high propagation latency.
- MEO satellites offer a balanced compromise between altitude and latency, making them suitable for regional communication services.
- LEO satellites, operating at lower altitudes, offer reduced latency and path loss, addressing the demands for high-bandwidth, low-latency applications.
- VLEO satellites, operating at ultralow altitudes, provide minimal latency and path loss, making them ideal for real-time communication and high-resolution Earth observation.

Table 2.1 Comparison of satellite features by orbit type.

Satellite feature	GEO satellites	MEO satellites	LEO satellites	VLEO satellites
Orbital altitude	35,786 km	2,000–35,786 km	200–2,000 km	100–200 km
Orbital period	24 h	2–8 hours	90–100 minutes	10–90 minutes
Path loss	High	High	Less	Least
Propagation latency	High	High	Low	Very low
Coverage	Largest	Large	Small	Very small
Satellite life	10–15 years	10–15 years	From a few years up to 10–15 years	A few months to a few years
Satellite required	At least 3	At least 6	Depends on the design	Requires a large constellation

It is essential to recognize that satellites operating in various orbital regimes exhibit a spectrum of advantages, disadvantages, and applications that are tailored to distinct objectives.

GEO satellites, positioned at an altitude of approximately 35,786 km, offer extensive coverage capabilities, with a single satellite able to encompass nearly one-third of the Earth's surface. Notwithstanding their reliability, GEO satellites experience a considerable signal transmission latency ranging from 250 to 300 ms. These attributes render them particularly suitable for large-scale broadcast communications and applications necessitating high levels of stability, such as television broadcasting and international data transmission for enterprises.

Moreover, MEO satellites, positioned at altitudes averaging between 2,000 and 35,786 km, demonstrate moderate signal latency, which varies between 50 and 150 ms. This particular attribute establishes a favorable balance between coverage and latency, thus necessitating a diminished quantity of satellites in contrast to LEO systems. The implementation of MEO satellites is especially advantageous for applications such as GPS, owing to their decreased system complexity and associated costs.

LEO satellites, which are distinguished by their relatively low operational altitudes ranging from 200 to 2,000 km, present numerous advantages. Firstly, the signal transmission latency associated with these satellites typically falls within the range of 10 ms, which is beneficial for facilitating real-time communication. Secondly, LEO satellites have demonstrated low terminal requirements and a high degree of deployment flexibility. Nevertheless, it is imperative to acknowledge that each satellite provides coverage over a limited geographical area and generally possesses a lifespan of approximately 5–10 years. Despite these constraints, LEO satellites are particularly well-adapted for real-time applications, including voice communication and video conferencing.

VLEO satellites, operating at altitudes ranging from 100 to 200 km, demonstrate remarkably minimal signal latency, specifically under 1 ms, alongside superior imaging and detection resolution pertaining to the Earth. Nonetheless, VLEO systems encounter a multitude of challenges, which encompass restricted coverage areas for individual satellites, diminished operational lifespans attributable to atmospheric drag, and the necessity for extensive satellite constellations to guarantee uninterrupted service.

2.1.2 Operational Modes

In the realm of the integration of satellites with 5G networks, two predominant operational modes have been delineated as frequently utilized, as illustrated in Figure 2.2: the transparent forwarding mode, which is alternatively denoted as the

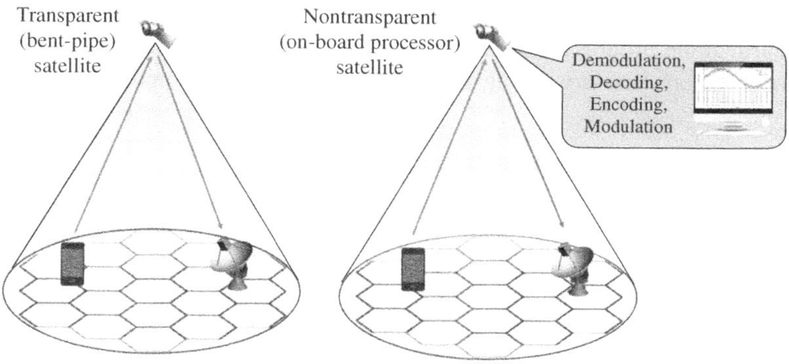

Figure 2.2 Two transmission methods for satellites on the NTN network.

bent-pipe mode, and the OBP mode. It is critical to emphasize that these two operational modes exhibit considerable divergence concerning the application scenarios and the inherent system complexity they engender.

The transparent forwarding mode operates as a fundamental signal relay architecture. In this mode, satellites are tasked with receiving, amplifying, and converting signal frequencies without engaging in any supplementary processing, thereby transmitting them directly to ground stations. This arrangement is characterized by its technical simplicity and cost-effectiveness, rendering it appropriate for extensive deployment, encompassing low-cost satellite constellations. Nevertheless, it exhibits a significant reliance on ground stations for signal processing, which may lead to increased latency and limit the overall flexibility of the system.

In contrast, the OBP mode empowers satellites to undertake both partial or comprehensive signal processing operations, which encompass demodulation, decoding, routing, and resource allocation, directly within the satellite system. By alleviating the burden of these operations from terrestrial ground stations, OBP mitigates transmission latency, augments spectral efficiency, and facilitates enhanced autonomy. As a result, OBP is particularly advantageous for applications demanding low latency and high bandwidth, such as instantaneous communication and sophisticated 5G services.

From a long-term perspective, the transference of various segments or the entirety of the functionalities of terrestrial stations to satellite systems signifies a pivotal trend in development. This methodology diminishes processing delays within the system and enhances user experience by facilitating a more seamless and efficient integration of satellite-terrestrial networks.

A more thorough and comprehensive discussion of the workflow, performance, and standardization progress associated with the two satellite operational modalities, namely transparent forwarding and OBP, will be elucidated in Chapter 9.

2.1.3 Inherent Characteristics and Unique Challenges

Although the development of SCSs is in full swing, they tend to suffer from numerous unprecedented challenges due to their inherent characteristics, which are clarified as follows:

1. **The Specific Orbit of Satellites Degrades Both the Security and Reliability:** LEO satellites are positioned between terrestrial communication systems and MEO satellites, rendering them particularly vulnerable to malicious interference. For instance, in comparison to MEO and GEO satellites, terrestrial assailants can effectuate equivalent jamming outcomes at low jamming power, attributable to the proximity of LEO satellites to Earth. In contrast, MEO and GEO satellites can also act as jammers to launch malicious jamming to contaminate the downlink signals of LEO satellites. Compounding the issue, the extensive coverage area associated with MEO and GEO satellites permits a singular satellite of either category to potentially disrupt multiple LEO satellites simultaneously. The phenomenon of spectrum congestion, resulting from the limited availability of radio resources, necessitates an unavoidable spectrum-sharing paradigm between SCSs and terrestrial communication systems. Furthermore, the specific spatial positioning of LEO satellites may precipitate co-channel interference (CCI) with both GEO SCSs and terrestrial communication infrastructures. Typically, a substantial number of LEO satellites are located in the intermediary space between terrestrial communication systems and MEO SCSs. Significant CCI may occur whenever LEO satellites traverse the line of sight (LoS) pathway of a GEO satellite within spectrum-sharing contexts [1].
2. **The High Mobility of Satellites Degrades Both the Security and Reliability:** The elevated mobility alongside the restricted coverage area of individual satellites leads to a diminutive duration spent above the horizon; consequently, the ground segment of SCSs should be responsible for the mobility management of terminals. Furthermore, to mitigate the risk of nefarious intrusions, the ground segment is typically permitted to authorize user access subsequent to a verification process [2]. Thus, the dual demands of mobility management and the authentication of a substantial number of terminals present considerable challenges for the ground segment. Additionally, in the absence of ISLs within the space segment, a multitude of gateways must be established to ensure the dependable operation of all satellites, which remain susceptible to cyberattacks. In a separate context, the substantial

rise in the quantity of satellites positioned in various orbits will inevitably heighten the likelihood of satellite collisions. Moreover, the escalation of launch activities has led to a notable increase in LEO space debris, which travels at high velocities and poses significant challenges to the reliable functioning of LEO satellites.

3. **The Large Number of Satellites or Gateways Degrades the Security:** SCSs must depend on a considerable array of gateways (transparent without ISLs, exemplified by OneWeb [3]) or a multitude of satellites (OBP with ISLs, for example, Iridium [4]) to realize comprehensive global coverage. The presence of a substantial number of LEO satellites or gateways engenders a plethora of opportunities, correspondingly.
4. **The Limited Resources of SCS Degrade Both the Security and Reliability:** Both the satellites and the terminals within the SCS possess restricted power, storage capabilities, and computational resources. On one hand, in an effort to minimize both the manufacturing and launch expenditures associated with satellites, the mass of a standard Starlink satellite is merely 227 kg, whereas the mass of a OneWeb satellite is under 150 kg [5]. These satellites are required to be outfitted with diminutive batteries, alongside constrained fuel and computational apparatus. In contrast to terminals situated in urban environments, those functioning in remote locales experience a deficiency of stable power sources, which must rely either on the battery they carry or on solar photovoltaic panels [6]. Consequently, the restricted computational and storage capabilities of LEO satellites render them ill-suited for intricate signal processing tasks. Furthermore, the terminals characterized by limited power availability yield low transmit power, which subsequently results in a diminished signal-to-noise ratio (SNR) for the signals transmitted from satellites, thereby presenting substantial obstacles to ensuring reliable signal reception. Specifically, to enhance the detection efficacy of faint signals under conditions of low SNR, the receiver must extend the coherent integration time, which necessitates the storage and processing of additional data [7].
5. **The Production of Low-cost Satellites Degrades Both the Security and Reliability:** The implementation of extensive satellite networks within multilayer mega-constellations has catalyzed a significant evolution in their production and testing methodologies. Consequently, a substantial quantity of low-specification components utilized in satellite manufacturing is procured from civilian suppliers, primarily to achieve cost efficiencies and to diminish the duration of the production cycle by utilizing commercial off-the-shelf (COTS) components. For example, OneWeb has established itself as a frontrunner in the mass production of satellites, with its manufacturing facility located in Florida anticipated to yield as many as two satellites daily [8].

Nevertheless, deficiencies in production techniques and insufficient testing protocols may result in potential malfunctions within the satellites. To accommodate a broader range of operational scenarios and facilitate iterative enhancements of functionalities, satellites are increasingly incorporating a substantial number of field programmable gate arrays (FPGAs), which are characterized by their adaptability and programmability. However, these FPGAs are also vulnerable to the detrimental effects of cosmic radiation [9], which can compromise the integrity of the programs and algorithms executing onboard the satellites.

6. **Quality of Service (QoS) Guarantee When Designing Security and Reliability Solutions:** A specified target QoS for satellite communication links. Satellite service providers are able to utilize these indicators to assess and enhance the operational efficiency of their systems, thereby ensuring that users consistently receive a superior QoS experience. In contrast to terrestrial communication infrastructures, the diverse range of services available is attributable to the extensive coverage provided by each satellite. With the advent of innovative services predicated on SCSs, there exists a necessity to deliver differentiated service capabilities. For instance, smart mining necessitates only low-rate data transmission accompanied by a substantial number of connections, whereas telemedicine demands exceptionally reliable, low-latency data transmission to facilitate teleconsultations and remote patient monitoring. Consequently, in the process of formulating solutions aimed at enhancing security and reliability, it is imperative also to establish QoS assurances for differentiated services.

2.1.4 Standardization Progress

The initiation of satellite launches occurred over five decades ago. The foundational idea of LEO SCSs can be traced back to the 1990s, during which Iridium [10], Globalstar [11], and Orbcomm [12] were conceptualized to deliver low-latency voice and data services. Nonetheless, several of these endeavors ultimately faced bankruptcy as a result of exorbitant expenditures, underdeveloped technology, and limited communication efficacy [13]. However, the advent of innovative materials, advanced technological methodologies, and economic scalability has heralded a new epoch for LEO SCS. In recent years, fueled by the escalating demands [14], diminished costs [8], and advancements in technology, LEO mega-constellations, including OneWeb, Starlink, and Lightspeed, are making concerted efforts to extend services to "the other 3 billion" individuals who presently lack Internet accessibility. At the current point of documentation, these systems appear to be evolving toward a unified architecture, as illustrated in Figure 2.3.

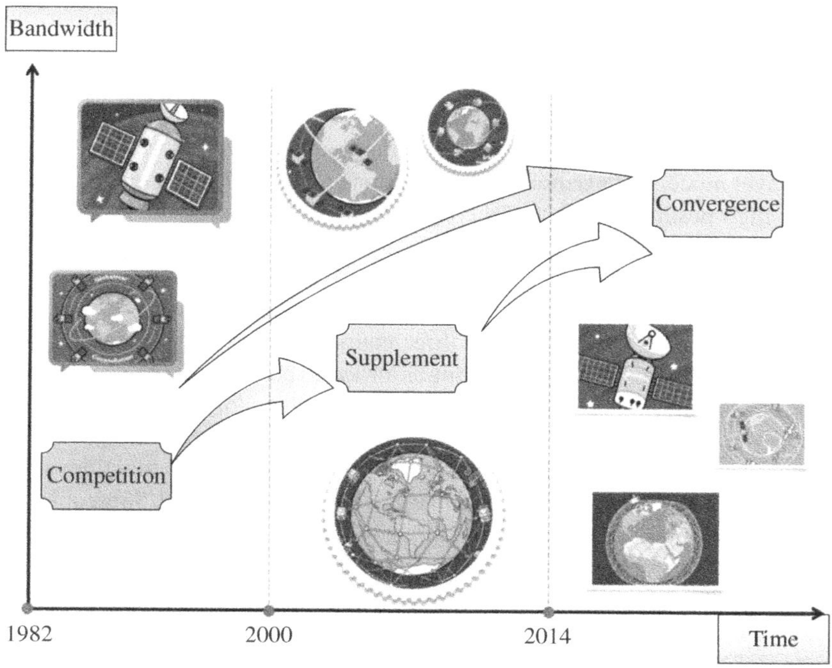

Figure 2.3 Development process of SCSs.

Meanwhile, 5G services have been rendered accessible to consumers equipped with smartphones; however, there exists a substantial demand among telecommunications operators to extend 5G services to massive machine type communication (mMTC) devices, particularly in underserved remote regions [15]. Consequently, both the academic research agenda and industrial initiatives are increasingly orienting toward NTN, epitomized by LEO satellites.

3GPP serves as a principal global organization tasked with the formulation of technical specifications pertinent to mobile wireless communication networks. The facilitation of 5G system to accommodate NTN necessitates a thorough and integrative design framework that encompasses a multitude of domains, with particular emphasis on radio access networks (RANs) and service aspects (SA).

1. **RANs:** The work on NTN started in 2017 with a Study Item in Release-15 in 3GPP RAN [16] focused on deployment scenarios and channel models for NTN. After completing this study, 3GPP followed up with Release-16, which aimed to provide solutions for adapting 5G new radio (NR) to support NTN. The main objective of this study is to identify a set of characteristics, including architecture, higher layer protocols, and physical layer components, that enable NTN within the 5G system while minimizing the

impact on the existing 5G system. The outcome of this study is summarized in [17]. The Release-17 Work Item focused on addressing the issues left from the Release-16 study, such as architecture, frequency synchronization, and hybrid automatic repeat request (HARQ). Specifically, two possible architectural options are discussed: transparent mode and onboard processing (regenerative) mode. Furthermore, user equipment (UE) with GPS capabilities can use their position and the NTN ephemeris to predict and compensate for the Doppler frequency shift [18].

2. **Service Aspects:** The 3GPP SA workgroup started to discuss the use cases of satellite-based NTN as part of the study item on satellite access in 5G based on Release-15. This study identified three main categories of use cases and service requirements, including serving as an access point for UE and a backhaul link, guaranteeing coverage for IoT devices, and supporting mission-critical access in disaster situations [19]. Based on this, the 3GPP SA workgroup further identified the main functions of the network, how these functions were related to each other, and the information they exchanged [20]. Subsequently, the scope of this work was to identify key issues and provide efficient solutions, such as mobility management, satellite delay, and QoS, which are captured in TR 28.808 [21]. Furthermore, the 3GPP workgroups approved the study on the architecture of the 5G core network [21] and the operation of IoT NTN [22]. The deployment of multilayer mega-constellations and work at 3GPP provide a potential pathway for the closer integration of terrestrial networks and NTN. At the time of writing, many researchers and commercial companies are tirelessly striving to integrate these two types of networks.

2.2 Ground Segment for Satellite Communication Systems

The ground segment of SCSs assumes a pivotal role in the comprehensive operation of the system. It is tasked with fundamental responsibilities, including data transmission, tracking, and control, as well as resource management between satellites and terrestrial networks. The proficient functioning of the ground segment is essential for the stability and efficacy of the entire system. The ground segment comprises gateways and the NCC, which operate in concert to guarantee dependable satellite communication and uninterrupted interaction with ancillary network systems. This coordination ensures the satisfaction of global communication requirements.

Gateways constitute a crucial component within the ground segment, undertaking the principal responsibility for the facilitation of uplink and downlink communication channels with satellites. In the operational framework, gateways acquire data from terrestrial users and subsequently transmit it to satellites through the uplink; concurrently, they receive data from satellites via the downlink and relay it to terrestrial networks or end users. Given the low orbital altitude of LEO satellites and the resulting dynamic orbital behavior, gateways are required to possess advanced tracking capabilities to sustain a stable communication link with the designated satellite; in addition, gateways are tasked with functions such as frequency conversion and power amplification during the transmission of signals, thereby guaranteeing optimized performance of the data link. The lack of ISLs necessitates an even greater dependence on gateways; in such scenarios, the implementation of gateways becomes increasingly critical. To facilitate communication with ground stations during each satellite's visibility period, it is imperative to ensure the deployment of a sufficient number of gateways on a global scale. These gateways must be interconnected through high-speed optical fiber networks, which provide advantages such as elevated bandwidth, minimal latency, and high reliability, thus effectively accommodating the extensive data transmission demands of SCSs. Moreover, the strategic selection of gateway deployment locations is contingent upon considerations such as satellite coverage, geographical factors, and fiber-optic availability, thereby ensuring the efficient operation of the system.

The NCC operates as the principal management and scheduling hub of the ground segment, executing essential functions such as satellite orbit manipulation, resource distribution, and fault management for the SCS. By employing real-time monitoring of satellite orbital positions and link statuses, the NCC facilitates enhanced satellite orbit management and resource distribution. Moreover, the NCC is responsible for the allocation of uplink and downlink frequencies and bandwidths, thereby optimizing the routing of data traffic and guaranteeing the effective operation of the network. Additionally, it possesses the capability to promptly identify network anomalies and execute corrective actions to ensure the persistence and dependability of communication services. The NCC also promotes interaction between the SCS and ancillary communication systems, including terrestrial mobile networks and wireless local area networks (WLANs). Through its gateways, the NCC conveys data obtained from satellites to the 5G/6G core network, thereby fostering uninterrupted communication between terrestrial and satellite users. Furthermore, the employment of satellites as WLANs access points has been shown to provide high-speed access services in remote or infrastructure-impoverished areas. This intersystem collaboration significantly augments the adaptability and scalability of the SCS.

In SCSs, ISLs possess the capability to facilitate direct data transmission between satellites, thereby diminishing reliance on terrestrial gateways. Nonetheless, in the absence of ISLs, the ground segment is compelled to depend on the establishment of multiple gateways to guarantee comprehensive communication coverage, which exacerbates the complexity of the design and configuration of the ground segment. Owing to the rapid orbital dynamics of each satellite and the ephemeral communication windows, a dynamic handover among various terrestrial gateways is necessitated to maintain link continuity. The establishment of a considerable number of gateways engenders significant financial expenditures and challenges in network management. These challenges may be mitigated by interconnecting gateways via high-speed optical fiber networks and optimizing data transmission pathways through redundancy and dynamic routing techniques, thereby minimizing transmission latency and enhancing network dependability. As ISL technology advances, inter-satellite data exchanges will further alleviate the demands placed on the ground segment, thereby enhancing the overall performance of the system.

The architecture of the ground segment encompasses a plethora of critical technologies, including satellite tracking, feeder link transmission, fiber-optic network interconnection, and network resource scheduling. In the realm of satellite tracking, it is crucial to utilize multibeam antennas in conjunction with automatic tracking systems deployed by ground gateways. This is necessary to accommodate the dynamic trajectories of the satellites and to ensure the stability of the ink as well as the integrity of real-time communication. Within the context of transmission via the feeder link, it is vital to optimize the signal quality at the gateway through meticulous frequency allocation and power management to mitigate losses associated with data transmission. The implementation of sophisticated scheduling algorithms and dynamic resource allocation frameworks within the NCC plays a pivotal role in facilitating an equitable distribution of network traffic. The consequent enhancement in system throughput and reliability stands as evidence of the effectiveness of these technologies functioning in concert to underpin the efficient operation of SCSs.

In conclusion, the ground segment assumes a crucial function in the formation of a linkage between the space segment and the terrestrial network within satellite communication frameworks. Through the synergistic operation of gateways alongside the NCC, the stability and efficacy of the system are preserved. In the absence of ISLs, the global deployment of gateways, interconnected via high-speed optical fiber networks, offers dependable support for satellite communication. Furthermore, through its interactions with terrestrial mobile communication systems and WLANs, the ground segment enables uninterrupted cross-network communication services. Technological advancements have prompted the transference of ground station functionalities to satellites, signifying a notable trend for future developments. This progression is anticipated to optimize the network architecture, diminish communication latency, and enhance the overall communication experience for global users.

2.3 User Segment for Satellite Communication Systems

The user segment represents a crucial element of SCS. Predominantly comprised of a substantial array of user terminals, this segment facilitates communication pathways with satellites via user links, which are accountable for the transmission of uplink data and the reception of downlink signals. The proficient functioning of the user segment exerts a significant influence on both the overall QoS and the user experience within the satellite communication framework. This segment assumes a pivotal role, especially in the realms of global communication coverage, mobile data transmission, and applications related to IoT.

The user demographic encompasses a diverse array of user terminals that can be systematically classified into three separate categories predicated upon their operational milieu and application contexts.

- **Fixed Terminals:** Fixed terminals are typically utilized in static locations such as ground stations or enterprise facilities, providing high stability and the capacity to access data at high throughput.
- **Mobile Terminals:** Mobile terminals, in contrast, are used in dynamic scenarios, such as on ships, aircraft, and vehicles, to support communication needs across regions or over oceans.
- **Portable Terminals:** Portable terminals are designed for individual users, including handheld satellite phones, mobile hotspot devices, and IoT sensors.

These terminals are characterized by their compact size, low power consumption, and convenience, making them well-suited for use in remote areas, disaster relief, and environmental monitoring. The variety of terminals within the system is such that they can be used to support a wide range of applications of SCSs.

In addition, the technical design of the user segment must meet various requirements, including communication frequency, terminal antenna design, and energy and power efficiency management. User links typically operate in frequency bands such as Ku, Ka, and L, each with its own advantages and drawbacks.

- The Ku-band offers a moderate frequency with good rain fade resistance, making it suitable for most environments.
- The Ka-band provides higher bandwidth and data rates but is more susceptible to weather conditions.
- The L-band, due to its strong penetration and low power requirements, is ideal for portable and low-power applications.

Moreover, the design of antennas and user terminals necessitates the incorporation of low-power, high-gain antennas that possess automatic tracking functionalities to effectively address the dynamic link challenges introduced by the rapid movement of satellites. Flat-panel antennas, owing to their compact dimensions and straightforward installation, are extensively utilized in mobile and portable applications, whereas high-performance parabolic antennas are more appropriate for fixed terminals, providing superior communication gain and reliability. Additionally, the management of power efficiency represents a considerable technical obstacle, particularly for portable terminals, which must sustain prolonged stable communication under conditions of limited power availability. Through the optimization of hardware architecture and energy consumption management algorithms, the operational lifespan of these devices can be significantly enhanced.

Consequently, the user segment constitutes an essential component in guaranteeing the efficacy of SCSs, as it caters to diverse communication requirements via an array of terminal types. The technical architecture of the user segment encompasses numerous elements, such as frequency allocation, antenna efficacy, and energy conservation, all of which are instrumental in facilitating resilient and efficient communication services for users across various contexts and applications.

2.4 Coverage Area

In light of the advancements in scientific and technological progress, mobile terminals are increasingly transcending their conventional deployment on terrestrial platforms and are now also being integrated into aerial vehicles. Moreover, due to the varying relative velocities in relation to satellites, the associated coverage technologies exhibit significant diversity. Typically, fixed beam and gaze beam technologies are employed to accommodate terrestrial users, whereas dynamic gaze beam technology is specifically utilized for unmanned aerial vehicles (UAVs) and high altitude platforms (HAPs), taking into consideration their operational speeds.

2.4.1 Ground Coverage Space

In the realm of satellite networks, the coverage model of satellites over terrestrial regions significantly impacts the system's efficiency and performance, particularly in the context of beam type selection. As illustrated in Figure 2.4, satellite systems are typically classified into two distinct coverage modalities, specifically fixed beam coverage and scanning beam coverage, predicated on the steerability of the satellite beams. These modalities present divergent advantages and challenges, profoundly affecting the network's handover performance, the continuity of service experienced by end-users, and the overall resource management of the satellite network.

The fixed beam coverage mode represents a satellite communication paradigm that employs a static directional beam, which does not undergo dynamic adjustments. This configuration permits the satellite to encompass a designated terrestrial area as it traverses its orbital path around the Earth. Given the swift movement

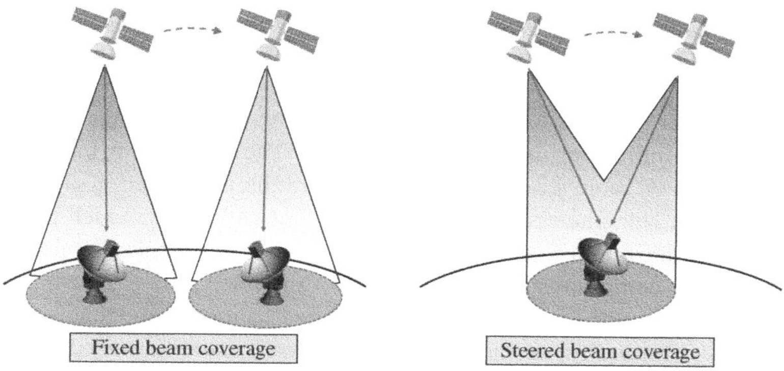

Figure 2.4 Satellite beam coverage modes.

of satellites, there exists a notable and rapid alteration in the coverage region. This circumstance necessitates frequent transitions between satellites, which transpire as they traverse over the Earth's surface. The elevated demand for the network's handover protocols stems from the imperative for uninterrupted service during these transitions. Notwithstanding the apparent straightforwardness of the fixed beam coverage mode, as demonstrated by its architecture and the lack of intricate hardware, such as that found in the Iridium constellation, it grapples with the challenge posed by frequent handovers. These transitions impose substantial signaling and computational burdens on the network, which may lead to a decline in service quality, especially for applications requiring continuous connectivity. Therefore, the effectiveness of handovers emerges as a crucial determinant in maintaining service continuity within such networks.

In contrast, the gazing beam coverage mode is distinguished by its adaptive modulation of the satellite's beam orientation, thereby ensuring persistent alignment with designated terrestrial cells. This mode enhances the provision of uniform coverage across a specified region for an extended timeframe, consequently diminishing the incidence of handovers and mitigating the potential for communication interruptions. The foremost advantage of this mode is that users are sustained under the auspices of a singular satellite for an elongated duration, thereby lessening the necessity for recurrent handovers and, in turn, augmenting service reliability. The gazing-beam coverage mode is particularly advantageous for applications necessitating uninterrupted connectivity and reduced latency, such as mobile broadband and HTS communication. The implementation of this mode empowers system architects to curtail the frequency of handovers, thereby decreasing signaling overhead and enhancing network efficacy. Nevertheless, it is imperative to acknowledge that the gazing beam coverage mode necessitates a more intricate technical infrastructure, specifically the integration of dynamic beam steering technology and superior hardware. This augmented technical complexity inevitably incurs elevated operational costs and heightened energy consumption.

It is apparent that both the fixed beam and the gazing beam coverage modalities possess their individual merits and drawbacks. The fixed beam modality is distinguished by its more straightforward configuration and superior satellite resource efficiency. Nonetheless, it is prone to frequent handovers, which place a considerable burden on the network's signaling and computational capabilities, thereby potentially compromising service quality. Conversely, the gazing beam modality provides extended user coverage, diminished handover frequency, and improved service continuity. Furthermore, it necessitates more advanced hardware and results in increased system expenditures. The choice between these two modalities is dependent upon the specific demands of the application and the availability of network resources, as each modality is tailored for particular satellite communication contexts.

2.4.2 Aeronautic Coverage Space

The dynamic gaze beam represents a highly sophisticated satellite communication and observational technology that is integral to contemporary satellite systems. This technology is predicated on state-of-the-art innovations such as PAs, whereby the phase and amplitude of each radiation element within the antenna array are swiftly and accurately modulated through electronic mechanisms. Consequently, the satellite is capable of dynamically modifying a range of beam parameters in real time, encompassing the direction, shape, width, and gain of the beam.

When the user operates within the aerial environment, exemplified by a drone in the HAP stratum, the advantages conferred by the satellite's dynamic gaze beam become distinctly apparent. In communication contexts, drones necessitate a heightened level of real-time responsiveness and stability owing to their incessantly shifting positions throughout their flight trajectories. Utilizing its dynamic gaze beam capabilities, the satellite is capable of swiftly capturing the positional data of the drone and precisely orienting the beam in alignment with the drone, thereby ensuring that the communication link between the two remains consistently stable and efficient. Irrespective of whether the UAV operates under intricate meteorological conditions or navigates frequently altering routes while executing its designated mission, the dynamic gaze beam can adeptly track its trajectory in real time, continuously providing dependable communication services for the UAV, and ensuring seamless and unobstructed data transmission between the UAV and the ground control center, as well as among multiple UAVs.

In the context of observational applications, the satellite's dynamic gaze beam enables it to rapidly direct its focus toward the geographical area occupied by the drone, whether it is facilitating the drone's reconnaissance of terrestrial targets, monitoring environmental fluctuations within a specific region, or guiding the drone in executing tasks such as search and rescue operations, it can achieve high-resolution imaging and precise data acquisition. Furthermore, in light of the intricate and variable airborne environment, which encompasses dynamic ionospheric interferences, frequent atmospheric condition fluctuations, and other adverse factors, the dynamic gaze beam demonstrates the capability to swiftly and adaptively modify its intrinsic characteristics according to prevailing conditions in order to optimize the quality of communication and observational outcomes. Concurrently, it also contributes to enhancing the spectrum resource utilization efficiency and the satellite system's communication capacity. By dynamically adjusting the coverage area and directional pointing of the beam, spectrum resources can be judiciously allocated to cater to the communication requirements of multiple aerial users, including drones, across diverse regions. In conclusion, the dynamic gaze beam of the satellite serves as a pivotal technical foundation for ensuring the efficient operation of UAVs within the aerial domain and for broadening their applicability, significantly augmenting the overall capability and application value of the satellite system within airborne operational contexts.

References

1 Qian Chen, Weixiao Meng, Shuai Han, et al. Service-oriented fair resource allocation and auction for civil aircrafts augmented space-air-ground integrated networks. *IEEE Trans. Veh. Technol.*, 69(11):13658–13672, Sep. 2020.

2 Biying Wang, Zheng Chang, Shancang Li, et al. An efficient and privacy-preserving blockchain-based authentication scheme for low earth orbit satellite assisted Internet of Things. *IEEE Trans. Aerosp. Electron. Syst.*, 58(6):5153–5164, Jun. 2022.

3 Di Zhou, Min Sheng, Jiandong Li, et al. Aerospace integrated networks innovation for empowering 6G: A survey and future challenges. *IEEE Commun. Surv. Tutor.*, 25(2):975–1019, Second Quart. 2023.

4 Stephen Pratt, Richard Raines, Carl Fossa, and Michael Temple. An operational and performance overview of the Iridium low earth orbit satellite system. *IEEE Commun. Surv. Tutor.*, 2(2):2–10, Second Quart. 1999.

5 Jinhui Huang and Jiang Cao. Recent development of commercial satellite communications systems. In *Artificial Intelligence in China*, pages 531–536. Springer Singapore, Feb. 2020.

6 Pietro Tedeschi, Savio Sciancalepore, and Roberto Di Pietro. Security in energy harvesting networks: A survey of current solutions and research challenges. *IEEE Commun. Surv. Tutor.*, 22(4):2658–2693, Fourth Quart. 2020.

7 David Gómez-Casco, José A López-Salcedo, and Gonzalo Seco-Granados. Optimal post-detection integration techniques for the reacquisition of weak GNSS signals. *IEEE Trans. Aerosp. Electron. Syst.*, 56(3):2302–2311, Jun. 2020.

8 Laurence Perino-Gallice, Olivier Masson, Michel Bel, and Florence Fusalba. Batteries for satellites constellation, using lean manufacturing for space industry. In *2019 European Space Power Conference (ESPC)*, pages 1–6, 2019.

9 Zhen Gao, Lingling Zhang, Ruishi Han, Pedro Reviriego, and Zhiqiang Li. Reliability evaluation of turbo decoders implemented on SRAM-FPGAs. In *2020 IEEE 38th VLSI Test Symposium (VTS)*, pages 1–6, 2020.

10 Zizhong Tan, Honglei Qin, Li Cong, et al. New method for positioning using Iridium satellite signals of opportunity. *IEEE Access*, 7:83412–83423, Jun. 2019.

11 Fred J Dietrich, Phil Metzen, and Paul Monte. The Globalstar cellular satellite system. *IEEE Trans. Antennas Propagat.*, 46(6):935–942, Jun. 1998.

12 Peng Wang, Jiaxin Zhang, Xing Zhang, Zhi Yan, Barry G Evans, and Wenbo Wang. Convergence of satellite and terrestrial networks: A comprehensive survey. *IEEE Access*, 8:5550–5588, Dec. 2019.

13 Roger Cochetti. *Low Earth orbit (LEO) mobile satellite communications systems.* In *Mobile Satellite Communications Handbook*, pages 119–156. Wiley, 2015.

14 Yvon Henri. *The OneWeb Satellite System.* In *Handbook of Small Satellites: Technology, Design, Manufacture, Applications, Economics and Regulation (Joseph N Pelton, and Scott Madry, editors)*, pages 1091–1100. Springer International Publishing, Cham, Feb. 2020.

15 Muhammad Asad Ullah, Konstantin Mikhaylov, and Hirley Alves. Massive machine-type communication and satellite integration for remote areas. *IEEE Wirel. Commun.*, 28:74–80, Aug. 2021.

16 T.38.811. Study on new radio (NR) to support non-terrestrial network. Oct. 2020.

17 T. 38.821. Solutions for NR to support non-terrestrial networks (NTN). Jan. 2020.

18 T. 23.737. Study on architecture aspects for using satellite access in 5G. Mar. 2021.

19 T. 22.822. Study on using satellite access in 5G. Aug. 2018.

20 T. 23.737. Study on architecture aspects for using satellite access in 5G. Jul. 2020.

21 T. 28.808. Study on management and orchestration aspects of integrated satellite components in a 5G network. Jan. 2021.

22 T. 24.821. Study on PLMN selection for satellite access. Feb. 2021.

Part II

Security Issues and Requirements

3

Satellite Security Issues

SCSs increasingly support civilian and military applications; therefore, it is crucial to eliminate the potential issues. In addition to the security threats posed by prospective adversaries, there exists a multitude of additional complications that are not attributable to the presence of aggressors, as illustrated by phenomena such as SEUs, collisions, and other relevant factors, particularly within LEO satellite networks. Accordingly, the concerns associated with satellite communications can be categorized into security attacks and reliability risks, and these two classifications further encompass their respective subclasses. Moreover, grounded in this categorization, this chapter further analyzes and infers the characteristics of these issues, including attribution, reversibility, awareness, and collateral damage. This chapter will proceed by emphasizing several insights gleaned from these challenges and utilizing them as a foundation to conceptualize prospective enhancements in security and reliability.

3.1 Introduction to Security Issues of Satellite Communications

SCSs, especially those operating in LEO, have become essential for an extensive array of civilian and military applications. Such systems facilitate worldwide connectivity, underpin critical infrastructure, and enable sophisticated technologies such as the Internet of Remote Things (IoRT) and 6G communication networks. Nevertheless, as dependency on LEO SCSs escalates, their vulnerability to a multitude of security threats concurrently intensifies. These threats can be broadly classified into two categories: security attacks and reliability risks. Security attacks represent intentional efforts by adversaries to undermine the integrity of the system, whereas reliability risks frequently arise from hardware malfunctions, environmental influences, or inherent design deficiencies. This subsection aims to

Safeguarding Satellite Communications: Issues, Challenges, and Solutions, First Edition.
Jianping An, et al.

delineate the various forms of security attacks and assess the degree of vulnerability of satellite systems to these emerging threats.

3.1.1 Various Types of Attacks

SCSs are vulnerable to a wide range of security attacks, which can be classified into two main categories: active and passive attacks. As shown in Figure 3.1, each category has different characteristics and poses unique challenges to the security and reliability of satellite systems.

- These assaults entail direct disruption of the satellite system, with the intent to hinder or harm its functionality. The most pressing and severe threats of this nature encompass node compromise and node destruction, where aggressors seize control of or physically obliterate satellite nodes, resulting in substantial collateral damage. Additionally, instances of active security attacks within physical layers encompass: (1) Power-based jamming [1], wherein high-energy signals are employed to overwhelm the satellite's receivers; (2) spoofing jamming [2], in which attackers replicate legitimate signals to mislead the authorized systems. Furthermore, active assaults can also manifest in the upper layers, including message alteration, wherein intercepted messages are modified to provoke erroneous decisions; DoS attacks [3], which inundate the system with excessive traffic, rendering it inaccessible to genuine users [4, 5].

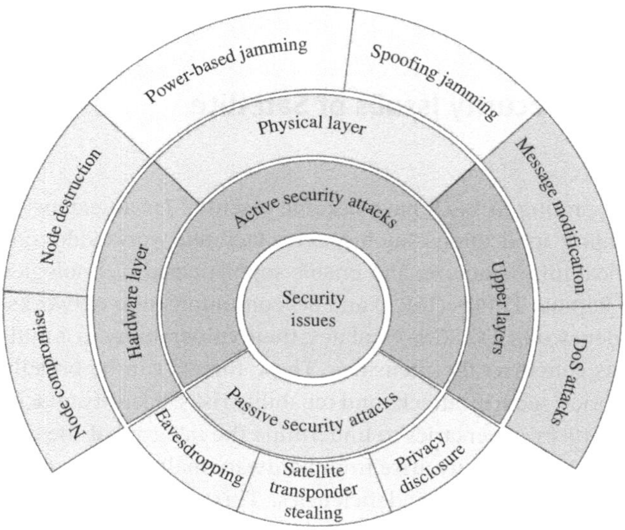

Figure 3.1 Classification of security attacks in satellite communications.

- Passive Security Attacks: In contrast to active attacks, passive attacks operate discreetly and aim to collect information without disrupting system operators. The most prevalent types of passive attacks consist of: (1) eavesdropping, where adversaries intercept and decode legitimate communications [6–8]; (2) hijacking satellite transponders, where criminals leverage existing satellite transponders to broadcast their own signals without authorization; (3) privacy breaches, where sensitive user information, such as location data, is disseminated across various satellites or gateways, potentially leading to misuse by unauthorized entities [9–11].

Active and passive attacks can lead to serious repercussions, including the compromise of sensitive data and the total interruption of satellite functions. Grasping these dangers is essential for creating robust counter-strategies to safeguard SCSs.

3.1.2 The Level of Exposure of Satellite Systems

The vulnerability of satellite systems to security risks is shaped by numerous factors, such as the inherent openness of wireless communication, the incorporation of COTS components, and the worldwide reach offered by satellites.

- **Open Nature of Wireless Communication:** The intrinsic openness of wireless communication renders satellite systems especially susceptible to interception and interference. For instance, the rapid mobility of LEO satellites and their regular access to user links present opportunities for eavesdroppers to capture signals using relatively straightforward equipment, such as drones [6–8]. Moreover, the substantial number of LEO satellites in orbit further complicates the challenge of safeguarding transmissions, as adversaries can take advantage of the system's extensive coverage to launch attacks from diverse locations.
- **Utilization of COTS Components:** In response to the increasing demand for LEO satellites, numerous manufacturers have opted for COTS components to minimize expenses and enhance production efficiency. Nevertheless, these components frequently lack the comprehensive security measures found in custom-designed hardware, rendering them vulnerable to potential attacks. For example, hardware trojan horses and microarchitectural side-channel attacks can jeopardize the overall system functionality by extracting sensitive information [12]. This dependence on COTS components creates considerable vulnerabilities, as malicious actors can take advantage of these flaws to gain unauthorized entry into satellite systems.
- **Global Coverage and Wide Reach:** The extensive global coverage offered by LEO satellites makes them appealing targets for adversaries. Attackers have the capability to execute jamming or spoofing assaults from various platforms in space, air, or on the ground, each presenting different levels

of intensity and scope. For instance, jamming from space can interfere with downlink signals across a large area, whereas ground-based jamming can effectively obstruct uplink signals with significant power and persistent disruption [1, 13]. The option to initiate attacks from diverse locations complicates the detection and management of these threats.

In conclusion, the vulnerability of satellite systems to security risks is heightened by the transparent characteristics of wireless communication, the incorporation of COTS components, and the worldwide reach afforded by satellites. These elements present a multitude of chances for adversaries to execute both active and passive assaults, underscoring the necessity for the establishment of strong security protocols to safeguard SCSs against harmful entities.

3.2 Satellite Security Attacks

SCSs function as a highly efficient and adaptable platform for a diverse array of military operations. Given their strategic significance, these systems frequently become primary targets for adversarial assaults. Their ground segments oversee all interactions with terrestrial communication networks, and these facilities can present opportunities for attackers. The susceptibility of SCSs is further intensified by their heavy dependence on ground segments, which manage every interaction with terrestrial communication systems. These ground-level infrastructures serve as vital hubs within the satellite communication framework, and their vulnerability to security threats opens up numerous avenues for adversaries to take advantage of security weaknesses.

In light of the nature of the attack, satellite security breaches can be further divided into active and passive security attacks, both of which are explained in detail in the following sections.

3.2.1 Active Satellite Security Attacks

In active security attacks, harmful actions are executed to interrupt or potentially harm the functioning of the system. Consequently, the targeted individual is made aware of these attacks. Active security attacks pose risks to both integrity and availability.

The most prevalent types of active security attacks consist of power-based jamming, spoofing jamming, message alteration, DoS, node compromise, and node destruction. Detection-oriented strategies are employed to reduce the effects of issues and expedite the remediation process [14].

3.2.1.1 Power-based Jamming

A straightforward approach to interfere with the reception of valid signals involves activating power-based jamming on the wireless user link of the SCS [1]. The majority of satellites currently in orbit employ the bent-pipe transponder, which functions without DSP. In this configuration, satellites merely amplify and shift the frequency of incoming signals from uplink to downlink before relaying them back to Earth, without engaging in demodulation or decoding, much like a bent pipe. As a result, these satellites are especially susceptible to power-based jamming assaults, as they lack the capability to effectively differentiate between authentic signals and interference.

Malicious actors can readily disrupt the operational integrity of the satellite by deploying high-intensity jamming signals. If the jamming intensity exceeds a critical threshold, it can potentially damage the receiver's front end of the satellite. Owing to the extensive open wireless link connecting satellites to Earth, adversaries have the capability to interfere with the signal from various locations, which can be categorized into the types depicted in Figure 3.2.

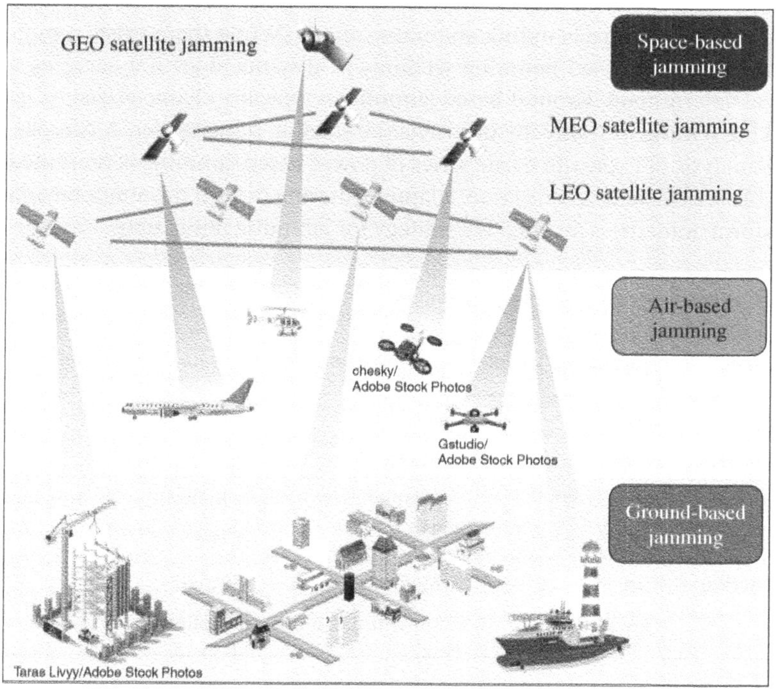

Figure 3.2 Sources of power-based and spoofing jamming contaminating SCSs.

- **Space-based Jamming:** Space-based jamming is predominantly emitted by spacecraft. This particular form of jamming possesses a substantial range within which it can interfere with the downlink transmission; however, it is constrained by limited jamming duration and power, owing to its restricted visibility above the horizon.
- **Air-based Jamming:** The opponent is capable of producing electronic jamming from aircraft or airships utilizing aerial platforms. Given that electronic jamming aircraft and airships possess greater maneuverability than their space-based equivalents, they are well-suited for deploying burst-type jamming techniques. In contrast to space-based jamming, the efficacy of air-based jamming tends to be significantly greater. As airships are typically positioned between terrestrial users and satellites, they have the potential to disrupt intended communication during both uplink and downlink transmissions.
- **Ground-based Jamming:** The efficacy of ground-based jamming is generally substantial, and its methods are varied, as this jamming is intentionally deployed by extensive fixed, vehicle-mounted, or maritime jamming stations equipped with significant resources and power. Ground-based jamming primarily impacts uplink communications. While there exists a multitude of ground-based jamming techniques, distance does not serve as a critical determinant. Ground-based jamming is typically characterized by its blocking nature, as it directly obstructs the satellite transponder. A comparative analysis of these three categories of power-based jamming is presented in Table 3.1. To address this issue, adaptive filtering in both the temporal and transform domains is an effective strategy for jamming mitigation.

Table 3.1 Comparison of power-based jamming.

Jamming types	Space-based jamming	Air-based jamming	Ground-based jamming
Jamming power	Low	Medium	High
Jamming time	Burst	Burst	Continuous
Resources	Limited	Limited	Rich
Mobility	Poor	Strong	Poor
Sphere of action	Large	Medium	Small
Scenarios	Downlink	Downlink Uplink	Uplink

3.2.1.2 Spoofing Jamming

Spoofing jamming constitutes a sophisticated form of electronic assault wherein the perpetrator deceives a receiver into accepting the authenticity of a malicious signal generated by the assailant. In contrast to power-based jamming, spoofing jamming demands a higher level of technical acumen. The assailant must possess a comprehensive understanding of the signal characteristics, including the waveform at the physical layer, the configuration of the frame, and other pertinent elements, to replicate its equivalent and mislead legitimate receivers.

A prevalent instance of spoofing jamming is observed in civilian GPS systems, where adversaries can effortlessly transmit counterfeit GPS signals to disseminate misleading information due to the widely known format of the civilian GPS signal [2]. In a manner analogous to GPS, there typically exists a designated downlink pilot channel utilized for the transmission of channel status, user management data, call details, and other pertinent information, as demonstrated by Iridium. Malicious actors have the capability to replicate the dedicated pilot channel to relay false information to legitimate users, thereby inducing network disruption.

Fortunately, there exist several established techniques to mitigate spoofing jamming, including energy detection, the utilization of multiple antennas, and authentication mechanisms. However, both energy detection and the deployment of multiple antennas tend to augment the complexity of the terminal apparatus. Consequently, the most efficient strategy is to implement authentication protocols for SCSs.

3.2.1.3 Message Modification

Message modification constitutes a category of cyber assault wherein a hostile entity intercepts, modifies, or manipulates transmitted communications, thereby undermining the integrity and reliability of the exchange. This form of attack can manifest in various ways, including message alteration, message insertion, and message deletion.

Message modification attacks are notably widespread within the ground segment, where malicious actors can unlawfully acquire data operation permissions and manipulate transmitted information. Given that the ground infrastructure is tasked with overseeing data processing, satellite control, and network coordination, an adversary's capacity to modify messages in this area represents a significant security threat. Altered control signals could lead to unauthorized satellite maneuvers, compromised navigation data, or interruptions in service. To mitigate the risk of message modification attacks, existing SCSs generally contemplate the implementation of intrusion detection systems (IDSs) and encryption algorithms [3].

3.2.1.4 DoS

A DoS attack constitutes a nefarious endeavor by an unauthorized individual to incapacitate a device or network, consequently making it unavailable to legitimate users. In SCSs, DoS attacks frequently aim at both the terrestrial segment and the orbital segment, significantly compromising the operational integrity of the network.

One of the most commonly employed techniques for executing a DoS attack entails inundating a network server with an excessive volume of traffic. In this particular instance, the perpetrator consistently dispatches requests to the targeted satellite. The satellite in question becomes preoccupied with addressing these illegitimate requests, thereby neglecting authorized users. Nevertheless, within the ground segment, which is tasked with the authentication of legitimate users, an attacker may impersonate an authorized user's identity to incessantly solicit authentication. This relentless surge of fraudulent authentication requests saturates the system, hindering its capacity to effectively manage genuine user access requests.

Beyond conventional DoS attacks, threat actors also leverage vulnerabilities or weaknesses in devices to execute a coordinated DoS assault on a singular target, commonly referred to as DDoS. The IoT botnet, from which malware source code has been disseminated, exemplifies a standard model of DDoS attacks in terrestrial IoT networks [4]. IoT botnets, established by cybercriminals, compromise a multitude of IoT devices and enlist them to conduct extensive DDoS offensives. Additionally, with the ongoing expansion of mega-constellations, the vast array of satellites increasingly presents a potential target for DDoS attacks. These assaults are particularly challenging to identify and counteract due to their utilization of hit-and-run strategies emanating from a multitude of IoT vectors distributed globally [15].

3.2.1.5 Node Compromise

A legitimate node can be compromised by an attacker utilizing malicious algorithms, programs, or software, thereby posing a significant risk to the integrity of the entire network. Once infiltrated, these nodes have the capability to inflict collateral damage through various mechanisms. For instance, these compromised nodes can intentionally divulge sensitive information to the attackers. Additionally, these nodes may deceive other legitimate nodes into becoming compromised themselves. Furthermore, an attacker could manipulate a compromised node to masquerade as a legitimate user or device, thereby usurping system resources. Detecting compromised nodes presents a considerable challenge, as the behaviors exhibited by these nodes can be indistinguishable from those of legitimate nodes. Implementing code patches is a widely adopted strategy to reduce the likelihood of such incidents occurring.

Furthermore, due to constraints in memory and processing capabilities, numerous satellites are unable to implement sophisticated encryption algorithms for their protection. By exploiting these security weaknesses, certain hackers may seize control of a satellite, effectively turning it into a compromised node by gaining access to its feeder link. Even more concerning, malicious actors could manipulate satellites to initiate self-destruction through harmful commands, or they may utilize specialized tools to deceive satellites, ultimately employing them to launch attacks on other satellites or space assets.

3.2.1.6 Node Destruction

The space segment, ground segment, and terminal devices associated with SCSs are susceptible to destruction and intentional assaults. In satellite-IoT applications facilitated by SCSs, power-constrained terminals, including oceanic buoys, which operate without sophisticated security protection protocols, may be vulnerable to capture by hostile entities. Moreover, satellites represent potential targets for anti-satellite weaponry, encompassing missiles and high-energy laser systems.

3.2.2 Passive Satellite Security Attacks

The most vital component of passive security breaches is that the targeted individual remains oblivious to the intrusion. Such passive security breaches may lead to a compromise of confidentiality. Three main categories of passive security breaches are examined, which include eavesdropping, satellite transponder stealing, and privacy disclosure.

3.2.2.1 Eavesdropping

The inherently open characteristics of wireless propagation render legitimate transmissions susceptible to the interception and analysis of signals or messages. Eavesdropping attacks necessitate minimal technical expertise. Individuals or corporate rivals can readily utilize a fleet of drones to capitalize on opportunities to monitor user connections due to the frequent access associated with the high mobility of satellites. Moreover, the extensive number of satellites facilitates eavesdropping due to the regular interactions and handovers required by these satellites. Once eavesdroppers successfully capture signals, they are capable of analyzing and extracting valuable information to facilitate future attacks. Direct-sequence spread spectrum (DSSS) and PLS techniques are employed independently to counteract eavesdropping [6].

Moreover, disruptions to standard operations or the unauthorized acquisition of confidential data may transpire during both the design phase and run-time due to hardware-related complications. To advance the discussion, in light of the remarkable increase in satellite deployments, numerous manufacturers are inclined to utilize COTS components to enhance their production efficiency while minimizing expenses. Nevertheless, certain COTS components may inadvertently create vulnerabilities that can be exploited by malicious entities. To address these potential threats, it is imperative to implement rigorous security evaluations and certification protocols for COTS components prior to their incorporation into satellite systems.

3.2.2.2 Satellite Transponder Stealing

Satellites, recognized for their expansive coverage and resilience against natural calamities, have garnered significant interest. Nevertheless, their prohibitive costs and sophisticated technology restrict manufacturing capabilities to a limited number of nations or organizations globally. Consequently, certain individuals lacking the capacity to produce satellites are taking advantage of existing satellite systems to execute illicit transmissions.

Existing satellites predominantly utilize OBP and transparent forwarding methodologies. Among these, transparent forwarding satellites are more susceptible to exploitation by malicious actors, as they do not engage in any form of signal processing. Consequently, it becomes infeasible to ascertain whether the data being received originates from a legitimate source. When perpetrators transmit their illicit signals, the satellite will continue to relay those signals [16].

To effectively exploit the satellite transponder discreetly, adversaries must engage in thorough research to ascertain the specific parameters of the satellite transponder, including but not limited to the operating frequency and orbital data of the satellite. Furthermore, DSSS techniques with a low power spectral density (PSD) are employed to obscure their presence within the legitimate frequency spectrum, as illustrated in Figure 3.3. Consistent modification of satellite operating parameters, such as the operating frequency, may serve as a deterrent against this form of intrusion.

3.2.2.3 Privacy Disclosure

LEO satellites possess the advantages of minimal path loss and reduced propagation delay. Consequently, they are well-equipped to facilitate access to the IoRT within the forthcoming 6G communication frameworks. Nonetheless, the dissemination and processing of satellite-based IoRT data could potentially result in a breach of user privacy.

The extensive data concerning registered UE must rely on a multitude of gateways (transparent without ISLs, e.g., OneWeb) or various satellites (OBP with ISLs, e.g., Iridium) for global exchange. Nevertheless, portions of this data may comprise sensitive information. For instance, to mitigate handover delays among satellites and facilitate high-precision, continuous user tracking, the location of UEs must be communicated between gateways or satellites [17]. Typically, access to

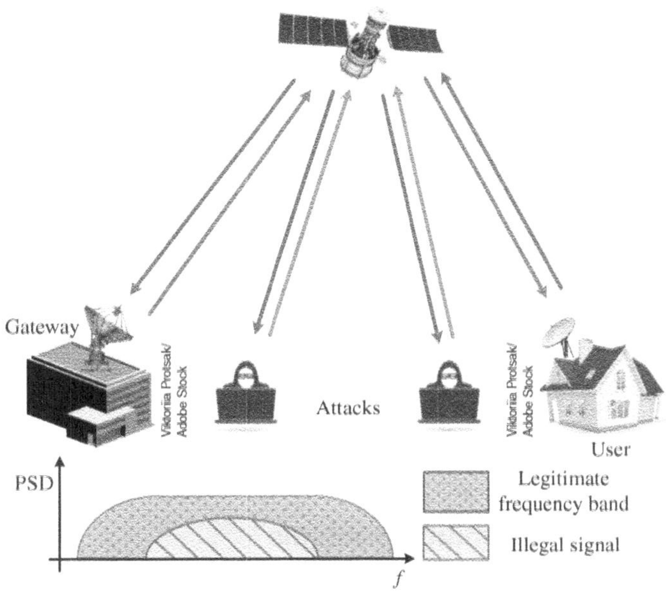

Figure 3.3 Illustration of transponder stealing.

the location of each UE is restricted to authorized personnel (e.g., designated service providers or insurance representatives). Once this data is disseminated, it falls outside the owner's jurisdiction and may be subject to unauthorized access, resulting in a significant risk of data exploitation. Furthermore, should an assailant gain access to user location data, this could instigate potentially irreversible attacks, including the possibility of physical harm.

Table 3.2 delineates the significant distinctions between active and passive security attacks. Passive security attacks, exemplified by eavesdropping, primarily seek to illicitly obtain sensitive information, such as passwords and communications. Conversely, active security attacks, which encompass message alteration, DoS, and node compromise, may be executed based on the intelligence gathered from passive security attacks. Perpetrators frequently leverage the confidential data acquired through passive eavesdropping to facilitate active security attacks. Furthermore, active security attacks can inflict considerable collateral damage when malicious nodes masquerade as legitimate entities and commandeer essential resources.

Table 3.2 The differences between the active and passive security attacks.

Characteristics	Passive security attacks	Active security attacks
Awareness	Not be aware	Aware
Against on	Confidentiality	Integrity as well as availability
Impact on system	No harm to system	System is damaged and its degree depends on the type of active attacks
Countermeasure	Prevention and mitigation	Detection and mitigation
Technical capacity	Simple to implement	Requires sophisticated technical capacities
Degree of difficulty to deal with	Easy to mitigate compared with active attacks	Tough to restrict

3.3 Security-related Reliability Problems

3.3.1 Intersystem Interference

Intersystem interference within LEO SCSs predominantly results from the sharing of spectrum resources between LEO SCSs and alternative systems, including GEO SCSs and terrestrial mobile communication networks [18]. The recent surge in the deployment of LEO satellites has significantly intensified the challenge of spectrum scarcity.

As the number of LEO satellites expands, the necessity for radio spectrum resources escalates. Nonetheless, the spectrum available remains constrained. LEO SCSs are required to operate alongside GEO SCSs within the identical spectrum band. This results in considerable CCI between the two systems. When LEO satellites navigate near the equator and function within the same frequency band as GEO satellites, they have the potential to generate heightened interference with the GEO satellites. In accordance with ITU regulations, LEO SCSs are classified as secondary users (SUs) and are obligated to avoid causing unacceptable interference to GEO SCSs, which are designated as primary users (PUs).

To alleviate this interference, a variety of methodologies have been examined. Power management is an essential strategy. In LEO SCSs, precise power management is necessary to adhere to the interference limitations imposed by GEO SCSs. For instance, scholars have conceptualized the power management dilemma as an optimization challenge, aiming to enhance the aggregate throughput of LEO SCSs while ensuring that the interference experienced by GEO SCSs remains within permissible thresholds [19]. Certain approaches utilize fractional programming techniques to reformulate the non-convex challenge into a more manageable format.

Moreover, beam steering strategies are also implemented. In specific instances, when an interference incident occurs, LEO satellites may deactivate the current beam and broaden its neighboring beam or modify the orientation of the spot beams to prevent disruption to GEO satellites. OneWeb has employed such strategies to minimize the potential for interference as its satellites traverse the equator.

Cognitive radio techniques are proficient in mitigating intersystem interference. Within cognitive radio networks, LEO SCSs, functioning as SUs, possess the capability to sense and monitor the surrounding spectral environment. They are able to identify the presence of GEO SCSs and modify their transmission power and operational parameters accordingly. For instance, certain studies have introduced spectrum sensing strategies that can ascertain the power levels of GEO SCSs, enabling LEO SCSs to adjust their transmission power to prevent interference. Furthermore, deep-learning-enhanced spectrum prediction techniques have been developed. Through the analysis of historical spectral data from GEO SCSs, these techniques can forecast future spectral occupancy, thereby aiding LEO SCSs in evading interference.

In the foreseeable future, as advancements in 6G systems progress and the integration of terrestrial systems, LEO SCSs, and GEO SCSs intensify, the repercussions of intersystem interference are anticipated to escalate significantly. Consequently, ongoing research and innovation in interference coordination methodologies are imperative to guarantee the reliable functioning of LEO SCSs and their harmonious coexistence with other systems.

3.3.2 Intrasystem Interference

Intrasystem interference within LEO SCSs primarily encompasses multiple access interference (MAI) and co-channel interference (CCI) among beams. Such interferences predominantly arise from the choice of physical waveforms and the limited availability of spectral information within the system [20].

MAI represents a prevalent challenge in spread spectrum (SS) systems deployed within LEO SCSs, particularly in military scenarios. While SS techniques offer a degree of immunity to interference, mitigating the near-far effect induced by MAI remains a significant challenge [21]. To alleviate this issue, power control and multiuser detection methodologies are routinely implemented. Power control facilitates the adjustment of transmission power among various users to harmonize the interference levels. Multiuser detection strategies enable the differentiation and segregation of signals from multiple users, thereby diminishing the adverse effects of MAI. Furthermore, the judicious selection of SS codes can contribute positively to this mitigation. Orthogonal complementary codes have demonstrated efficacy in attenuating MAI to a certain degree. Nevertheless, these orthogonal codes exhibit sensitivity to frequency shifts, necessitating additional research to resolve this vulnerability.

CCI between beams transpires when multi-beam satellites utilize shared frequencies within their operational domains. When disparate beams operate on identical frequencies, interference leakage is likely to occur in overlapping regions, particularly among adjacent beams. The angular sidelobes inherent in the beam radiation patterns play a significant role in this interference. To augment system capacity while mitigating this interference, transmit precoding (TPC) methodologies grounded in transmitter-side channel state information (CSI) may be utilized. Certain satellite systems have developed hybrid wide-spot beam configurations [22]. The spaceborne payload produces multiple fixed wide beams for extensive area coverage, thereby enhancing frequency reuse distance, and employs high-gain spot beams in areas of high traffic to boost capacity.

In contrast to terrestrial communication systems, LEO satellites offer an expansive coverage area accompanied by intricate service requirements. The design of the frequency reuse pattern must be meticulously crafted, taking into account the varying access demands across different geographical regions. In locations characterized by elevated access requirements, such as airports or railway stations, an appropriate frequency reuse strategy can be implemented to adequately fulfill the demand. Conversely, in regions with diminished access requirements, such as deserts or oceans, it is imperative to modify the frequency reuse pattern to mitigate CCI between beams. The dynamic and unbalanced nature of services resulting from the satellite's orbital movement presents a considerable challenge to the effective management of frequency reuse. A comprehensive overview of the distinctions and interrelations between intersystem interactions and intrasystem interference has been compiled in Table 3.3.

Table 3.3 Differences and connections between the two types of interferences.

Differences and connections	Intersystem interference	Intrasystem interference
Source	Spectrum sharing between LEO, GEO and terrestrial systems	Waveform selection within the system and spectrum scarcity
Types and causes	CCI occurs when operating on the same frequency as GEO, and LEO must not interfere with GEO	MAI causes the near-far effect; CCI occurs in overlapping areas due to beam frequency reuse
Solutions	Power control, beam drifting, cognitive radio technology	Power control, multiuser detection, selection of SS codes, TPC technology, design of beam architecture

Differences and connections	Intersystem interference	Intrasystem interference
Future development	The impact will be more severe with the development of 6G and system integration, and innovative interference coordination technologies are needed	Wide coverage, complex requirements, and satellite movement challenges frequency reuse control

3.3.3 Single-event Upsets

SEUs represent a significant reliability concern for LEO satellites. These events arise from the ionization effects of cosmic radiation particles, which result in the generation of numerous electrons and holes along the trajectory of the incident radiation. When these charge carriers are accumulated by electronic components, such as FPGAs utilized in the satellite, they may induce transient faults. Should the accumulated charge surpass the maximum threshold that the device can tolerate without experiencing SEUs, the logical state of the circuitry will be inverted [23].

The likelihood of SEUs is intricately connected to both the altitude and inclination of a satellite's orbit. Empirical studies indicate that in LEOs, at altitudes beneath 2,000 km, a higher orbital altitude correlates with an increased probability of SEU occurrences. Furthermore, a closer alignment of the orbital inclination to 90° significantly enhances the likelihood of SEUs [24]. Various electronic components exhibit differing levels of resilience against SEUs. Application-specific integrated circuits (ASICs) typically demonstrate superior resilience compared to FPGAs, but lack the flexibility of FPGAs. Owing to their adaptability and high performance, FPGAs find extensive application in LEO satellites; however, it is imperative to implement specific precautions to guarantee their dependable operation within the space environment.

To alleviate the effects of SEUs, a variety of techniques are utilized. One prevalent approach is triple module redundancy (TMR). In TMR, three identical replicas of the original program module are employed, with their outputs directed to a majority voter. In the event that one of the modules fails, the other two can conceal the fault, thereby ensuring the uninterrupted functionality of the system. Nonetheless, TMR requires threefold the resources of the original program module, which typically restricts its application to critical components of the program, such as the control section. Another approach is periodic refreshing. This technique can rectify errors by refreshing the program while maintaining its execution flow. However, when re-accessing the block random access memory (RAM) utilized in FPGAs, the real-time state may be compromised; consequently, the block RAM must also implement TMR to counteract the effects of SEUs.

In the conceptualization and functioning of LEO satellites, it is imperative to take into account the potential ramifications of SEUs. It is essential to identify and implement suitable SEU prevention and mitigation strategies tailored to the distinct requirements and resource limitations of the satellite, thereby ensuring the dependable operation of the satellite's electronic systems and the overall success of the mission.

References

1 Hossein Pirayesh and Huacheng Zeng. Jamming attacks and anti-jamming strategies in wireless networks: A comprehensive survey. *IEEE Commun. Surv. Tutor.*, 24(2):767–809, Second Quart. 2022.

2 Daojing He, Sammy Chan, and Mohsen Guizani. Communication security of unmanned aerial vehicles. *IEEE Wirel. Commun.*, 24(4):134–139, Aug. 2017.

3 Pradeepkumar Bhale, Debanjan Roy Chowdhury, Santosh Biswas, et al. OPTIMIST: Lightweight and transparent IDS with optimum placement strategy to mitigate mixed-rate DDoS attacks in IoT networks. *IEEE Internet Things J.*, 10(10):8357–8370, May 2023.

4 Graham Cluley. Could this be the world's most harmless IoT botnet? (May 08, 2020).

5 Ayan Roy-Chowdhury, John Baras, Michael Hadjitheodosiou, et al. Security issues in hybrid networks with a satellite component. *IEEE Wirel. Commun.*, 12(6):50–61, Dec. 2005.

6 Fuwang Dong, Wei Wang, Xin Li, et al. Joint beamforming design for dual-functional MIMO radar and communication systems guaranteeing physical layer security. *IEEE Trans. Green Commun. Netw.*, 7(1):537–549, Mar. 2023.

7 Zhiwei Lu and Yiwen Jiao. Efficiently all-digital code tracking for band-limited DSSS systems. *IEEE Commun. Lett.*, 27(2):686–690, Feb. 2023.

8 Fengyu Wang, Wei Cui, and Jing Tian. A super-resolution multipath estimation algorithm for DSSS systems. *IEEE Trans. Aerosp. Electron. Syst.*, 59(1):109–124, Feb. 2023.

9 Ruyan Wang, Shiqi Zhang, Boran Yang, Zhigang Yang, Puning Zhang, and Dapeng Wu. Enabling data sharing through data trusts in LEO satellite internet. *IEEE Wirel. Commun.*, 31(1):70–76, 2024.

10 Yong Zuo, Mingyang Yue, Mingchen Zhang, Sixian Li, Shaojie Ni, and Xiaojun Yuan. OFDM-based massive connectivity for LEO satellite internet of things. *IEEE Trans. Wirel. Commun.*, 22(11):8244–8258, 2023.

11 Zhisheng Yin, Nan Cheng, Yilong Hui, et al. Multi-domain resource multiplexing based secure transmission for satellite-assisted IoT: AO-SCA approach. *IEEE Trans. Wirel. Commun.*, 22(11):7319–7330, 2023.

12 Luca Cassano, Stefano Di Mascio, Alessandro Palumbo, et al. Is RISC-V ready for space? A security perspective. In *2022 IEEE International Symposium on Defect and Fault Tolerance in VLSI and Nanotechnology Systems (DFT)*, pages 1–6, Austin, TX, USA, 19–21 Oct. 2022.

13 Yulong Zou, Jia Zhu, Xianbin Wang, and Lajos Hanzo. A survey on wireless security: Technical challenges, recent advances, and future trends. *Proc. IEEE*, 104(9):1727–1765, Sep. 2016.

14 Cong Chen, Zhongyuan Jiang, and Jianfeng Ma. Privacy protection for marginal-sensitive community individuals against adversarial community detection attacks. *IEEE Trans. Comput. Social Syst.*, 11(1):782–794, 2024.

15 Rana Faisal Hayat, Sana Aurangzeb, Muhammad Aleem, Gautam Srivastava, and Jerry Chun-Wei Lin. ML-DDoS: A blockchain-based multilevel DDoS mitigation mechanism for IoT environments. *IEEE Trans. Eng. Manage.*, 71:12605–12618, 2024.

16 James Pavur, Daniel Moser, Vincent Lenders, et al. Secrets in the sky: On privacy and infrastructure security in DVB-S satellite broadband. In *Proc. Conference on Security and Privacy in Wireless and Mobile Networks*, pages 277–284, Miami, FL, USA, May 2019.

17 Huanxi Cui, Jun Zhang, Yuhui Geng, et al. Space-air-ground integrated network (SAGIN) for 6G: Requirements, architecture and challenges. *China Commun.*, 19(2):90–108, Feb. 2022.

18 Chi Zhang, Chunxiao Jiang, Jin Jin, Sheng Wu, Linling Kuang, and Song Guo. Spectrum sensing and recognition in satellite systems. *IEEE Trans. Veh. Technol.*, 68(3):2502–2516, 2019.

19 Yunfeng Wang, Xiaojin Ding, and Gengxin Zhang. A novel dynamic spectrum-sharing method for GEO and LEO satellite networks. *IEEE Access*, 8:147895–147906, 2020.

20 Jihwan Choi and Changhee Joo. Challenges for efficient and seamless space-terrestrial heterogeneous networks. *IEEE Commun. Mag.*, 53(5):156–162, 2015.

21 Fred Dietrich, Phil Metzen, and Paul Monte. The globalstar cellular satellite system. *IEEE Trans. Antennas Propag.*, 46(6):935–942, 1998.

22 Yongtao Su, Yaoqi Liu, Yiqing Zhou, Jinhong Yuan, Huan Cao, and Jinglin Shi. Broadband LEO satellite communications: Architectures and key technologies. *IEEE Wirel. Commun.*, 26(2):55–61, 2019.

23 Vanessa Vargas, Pablo Ramos, Raoul Velazco, Jean-François Mehaut, and Nacer-Eddine Zergainoh. Evaluating SEU fault-injection on parallel applications implemented on multicore processors. In *2015 IEEE 6th Latin American Symposium on Circuits & Systems (LASCAS)*, pages 1–4, 2015.

24 Fuqiang Zhang, Gang Guo, Yingcan Qin, et al. Prediction of proton-induced single event effect on SRAMs in-orbit soft error rate on typical satellite orbit. *Spacecr. Environ. Eng.*, 35(4):365–370, 2018.

4

Security Requirements

Satellite communication holds paramount importance across military, governmental, and commercial domains, necessitating the implementation of stringent security protocols to safeguard sensitive data. Essential security prerequisites encompass confidentiality, integrity, accountability, and security-related reliability, which incorporates factors such as reliability, latency, and throughput. These elements collectively ensure a secure and dependable communication system.

4.1 Confidentiality Performance

For contemporary satellite communications, the performance of confidentiality is paramount to safeguard the integrity of information. Serving as a means of long-distance, global communication, satellite communications are extensively utilized across various sectors, including military, governmental, and commercial domains, and encompass a significant volume of sensitive data. To shield such information from eavesdropping, tampering, or unauthorized access, satellite communications must maintain an elevated standard of confidentiality.

In satellite communications, the assurance of confidentiality performance is imperative to safeguard sensitive data from unauthorized access. Nonetheless, this constitutes merely one facet of the overall security framework. The reliability of the communication itself holds equal importance, significantly impacted by outage probability (OP) [1]. The OP denotes the probability of a communication disruption resulting from signal fading, interference, or various external influences. Within the realm of satellite communications, the OP becomes particularly critical during severe weather events, such as intense snowstorms or strong winds, which can lead to signal degradation and unstable connections. To bolster system stability and reliability, it is essential to mitigate the OP through a range of design

Safeguarding Satellite Communications: Issues, Challenges, and Solutions, First Edition.
Jianping An, et al.

optimizations. These strategies encompass the careful selection of satellite orbits, enhancement of the ground station's reception capabilities, and the selection of the most suitable frequency band. Such initiatives ensure a more robust communication system, capable of sustaining performance even in adverse environmental conditions.

In addition to general OP, in the context of secure communications, another important metric is the secrecy outage probability (SOP). The SOP is the likelihood that a communication outage will occur due to an attack or malicious interference that threatens the confidentiality of the communication process. Such outages are usually caused by network attacks, signal interference (e.g., jamming or blocking attacks), and malicious replay attacks. In order to reduce the probability of a security outage, SCSs need to adopt advanced encryption algorithms, authentication mechanisms, and anti-jamming techniques to ensure that the security and reliability of communications can be guaranteed even in the face of attacks. In fact, the system cannot guarantee perfect and secure transmission when only partial CSI is known [2]. Therefore, in order to portray the probability of failure of a complete secure transmission, researchers have proposed the concept of security OP. Currently, there are two types of definitions commonly seen for security OP. The first type of security OP is defined as the probability when the security capacity C_S is lower than the target security rate R'_S. The mathematical form is expressed as

$$P_{SO} \triangleq \mathbb{P}\{C_S < R'_S\}, \tag{4.1}$$

where $\mathbb{P}$ denotes the probabilistic operation; $C_S < R'_S$ event is referred to as a security outage event [3]. The probability of security outage under this definition includes two cases: one is that the rates of both the main channel and the eavesdropping channel can satisfy the requirements of the destination node and the eavesdropping node to decode the received information normally; the other is that the security capacity is too low due to the low channel rate of the main channel. However, at this time, since the destination node will not be able to decode the received information correctly, usually the system will interrupt the information transmission (connection outage event) [4]. In addition, for the second case where the security outage event occurs, the channel rate of the eavesdropping channel may also be very low and thus unable to decode the received information. It can be seen that $C_S < R'_S$ indicates that the system is unable to meet the requirement of secure transmission with insufficient channel and information collection, and further improvement of channel estimation is usually required. The second type of SOP is expressed through the following equation [5]:

$$P_{SO} \triangleq 1 - \mathbb{P}\{C_D \geq R'_D, C_S \geq R'_S\}, \tag{4.2}$$

where R'_D denotes the target rate of correct decoding by nodes that satisfy the target.

The second type of SOP is defined as

$$P_{SO} \triangleq \mathbb{P}\{R_D - R_S < C_E \mid \text{information transmission}\}, \tag{4.3}$$

where R_D and R_S denote the two rates set by the system to achieve secure transmission, i.e., the decoding rate and the security rate, respectively, and $R_e \triangleq R_D - R_S$ is defined as the cost of protecting the secure transmission of information from eavesdropping.

When the channel capacity C_D of the main channel is greater than R_D, it indicates that the destination node can decode the information correctly. When the channel capacity, C_E, of the eavesdropping channel is greater than R_e, it indicates that the perfect security of the information transmission cannot be guaranteed [6]. Obviously, the second type of security OP is premised on the fact that the information is inevitably transmitted, so it directly reflects the situation that the information cannot be transmitted perfectly securely. When it is assumed that the source node is unaware of the existence of the eavesdropper and will keep sending information, the second type of security outage definition can be rewritten as [7]

$$P_{SO} = \mathbb{P}\{R_D - R_S < C_E\}. \tag{4.4}$$

It has been studied for broadcast channels as well as nondegenerate eavesdropping channels to give the maximum error-free transmission of the message codeword rate, i.e., the security capacity, which guarantees the perfect security condition of the message, denoted as

$$C_S = \max_{V \to X \to YZ} \{I(V;Y) - I(V;Z)\}, \tag{4.5}$$

where $I(\cdot;\cdot)$ denotes the mutual information, and the auxiliary variable V satisfies the Markov chain $V \to X \to YZ$. There is no uniform way to select V, but its selection affects the maximum difference between two mutual information.

In the Gaussian channels, the safe capacity can be further calculated by [8]

$$C_S = [C_D - C_E]^+ = \left[\log_2\left(1 + \frac{P}{\sigma_D^2}\right) - \log_2\left(1 + \frac{P}{\sigma_E^2}\right)\right]^+, \tag{4.6}$$

where P is the transmit power of the source node, and σ_D^2 and σ_E^2 are the Gaussian white noise power at the destination and eavesdropping nodes, respectively. C_D and C_E denote the channel capacity of the main and eavesdropping channels, respectively.

The aforementioned expression for security capacity does not take into account channel fading, which is an inherent characteristic of wireless channels in real-world scenarios, including multipath propagation induced by airborne particulates, obstructions casting shadows, and precipitation in the atmosphere. When incorporating the effects of wireless channel fading, the security capacity may be articulated as

$$C_S = \left[\log_2\left(1 + \frac{P|h_D|^2}{\sigma_D^2}\right) - \log_2\left(1 + \frac{P|h_E|^2}{\sigma_E^2}\right)\right]^+, \tag{4.7}$$

where $|h_D|^2$ and $|h_E|^2$ represent the channel gain of the main channel and the eavesdropping channel, respectively [9].

4.2 Integrity Performance

Integrity holds a position of utmost significance within the realm of satellite communication, particularly regarding the transmission of sensitive information. It guarantees that data remains secure and intact throughout the transmission process. Integrity involves not merely the accuracy of the information but also its completeness, indicating that all elements of the information must stay unchanged during transit. To ensure integrity, especially when faced with a variety of potential attacks and disruptions, the deployment of effective security measures is essential.

In scenarios where a robust adversary disrupts a wireless communication link, the integrity of signal transmission may be distorted or wholly compromised through the introduction of high-power white noise across the entire frequency spectrum, resulting in a condition known as "wideband interference" or "signal contamination." The situation is further exacerbated in the case of LEO satellites due to their lower altitude, which allows attackers to effectively contaminate the signal with lower interference power compared to GEO satellites, which, being positioned much farther from the ground, are less susceptible to such jamming, as lower power is required to disrupt signals.

To address this challenge, some techniques have been proposed as a viable solution in Figure 4.1, including:

- **The DSSS:** DSSS is an SS communication technique. It multiplies the original signal with a high-rate pseudo-random code (PR code) to extend the spectrum of the signal to a very wide bandwidth. For example, the original signal bandwidth may be only a few hundred kilohertz, but after DSSS processing, the signal bandwidth may be extended to several megahertz or even tens of megahertz.

- **The frequency-hopping spread spectrum (FHSS):** FHSS is an SS communication technique that transmits signals by rapidly hopping between multiple frequency channels according to a certain pattern and rate. For example, an FHSS system may hop frequencies hundreds or even thousands of times per second, and the hopping frequency sequence is controlled by PR codes.

The more detailed differences and connections between DSSS and FHSS are shown in Table 4.1. The DSSS technique functions by spreading the signal over a more extensive frequency range using a spreading code [10]. This approach guarantees that even with reduced signal power, the signal is distributed across a wide frequency band, thus significantly enhancing its resistance to interference. The DSSS methodology is distinguished by its capacity to enable accurate signal demodulation even amid substantial interference, thereby ensuring the recovery of the original information. The intrinsic jamming resistance features of the DSSS methodology render it a superior option for countering low-power interference, such as white noise.

In addition to the application of DSSS, there are alternative anti-jamming strategies that can be utilized to further enhance communication integrity, including the implementation of FHSS, which intermittently modifies the carrier frequency to evade persistent interference associated with a fixed frequency, as well as the adoption of adaptive modulation and coding (AMC), a methodology that dynamically modifies the modulation and coding schemes based on channel conditions, thereby ensuring optimal communication quality even in challenging environments. The application of DSSS in satellite systems is critically important,

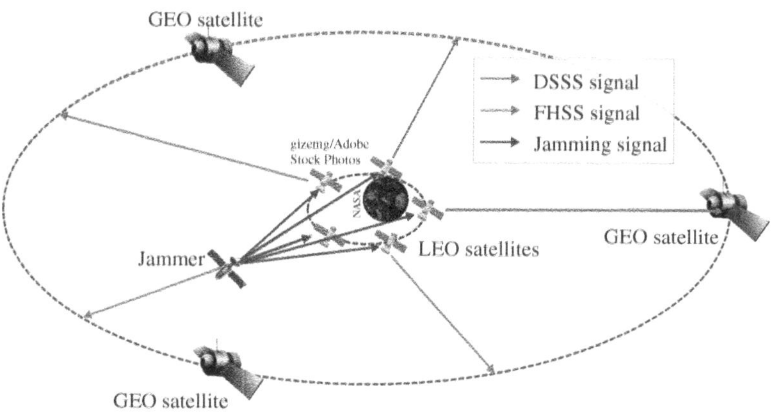

Figure 4.1 The analysis of DSSS and FHSS in multilayer satellites.

Table 4.1 Comparison of DSSS and FHSS: key characteristics and applications.

Comparison items	DSSS	FHSS
Anti-interference	Strong resistance to narrowband and multipath interference via spectrum spreading	Mitigates narrowband interference through frequent frequency hopping
Data rate	High (up to 11 Mbps or more)	Low (typically 1–2 Mbps in IEEE 802.11)
Bandwidth usage	Wide bandwidth (26 MHz) with low efficiency	Narrower bandwidth (1 MHz) with higher efficiency
concealment	Even spectral distribution enhances security	Frequency hopping reduces interception risk but is slightly less secure than DSSS
Complexity and Cost	Fully digital architecture increases complexity and cost	Traditionally high in cost and complexity, but digitalization improves efficiency
Applications	Mobile communication, GPS, satellite systems, and IoT requiring high data rates and multipath resistance	Wi-Fi, Bluetooth, LoRa, and ZigBee, favoring cost-sensitive and mobile scenarios
Anti-fading	Strong resistance by distributing information across frequency and time domains	Weaker resistance, vulnerable to frequency-selective fading and burst errors

considering the system's lower altitude and minimized communication latency. The utilization of this technique not only enhances the system's resilience against interference but also sustains high transmission reliability under adverse conditions, thereby protecting the integrity of the transmitted data. As advancements in technology progress, it is expected that modern satellite systems will integrate encryption and authentication techniques alongside DSSS, thereby further strengthening data security and integrity. The importance of integrity performance in the realm of satellite communication, particularly within satellite systems, is unequivocal. The implementation of methodologies such as DSSS enables effective counteraction against interference from adversaries, ensuring precise transmission of communication data and safeguarding the integrity of information. This, in turn, assures the security and reliability of satellite communication.

Furthermore, even in the absence of jamming, the Doppler frequency shift can exert a detrimental influence on the integrity of synthetic aperture radar systems, attributable to the high velocity and rapid motion of satellites in relation to the Earth's surface. This phenomenon arises from the relative dynamics between the satellite and the ground station, leading to a frequency shift in the transmitted signal due to the swift movement of the satellites. The magnitude of this shift can vary between 7 and 8 km per second. The implication of the Doppler shift is an erroneous alteration in the perceived frequency at the ground station, and empirical research has evidenced that this results in inaccuracies in signal demodulation and decoding. The extent of the Doppler shift is dependent on various factors, including the relative speed of the satellite with respect to the receiver, the altitude of the satellite, and the positional relationship of the ground station concerning the satellite's orbit. Consequently, there exists a potential for the signal received by the ground station to deviate from the expected frequency. This, in turn, possesses the potential to induce distortions or misalignments within the signal. In scenarios involving high-data-rate communications or applications that demand low latency, such discrepancies can lead to significant consequences. They have the potential to produce errors in data transmission, thereby impacting the precision and dependability of the transmitted information.

Therefore, in order to address the challenges associated with the positioning and velocity of satellites, SCSs utilize a range of sophisticated techniques, including algorithms for Doppler shift compensation. These algorithms facilitate dynamic adjustments to the receiver's frequency tuning, informed by real-time tracking of the satellite's positional and velocity parameters. Such adjustments are essential to uphold the integrity of the communication link. Furthermore, advanced signal processing methodologies, including frequency tracking and adaptive modulation schemes, are implemented to counteract the impacts of Doppler shifts and ensure that the received signal remains in sync with the transmitted signal. Nonetheless, despite these strategic initiatives, the considerable mobility of satellites presents an ongoing challenge to achieving the precise synchronization necessary for maintaining the integrity of the communication system, a concern that is particularly pronounced in systems necessitating high levels of precision and reliability.

In conclusion, although Doppler frequency shifts are not induced by external interference or disruption, they pose a considerable challenge to the maintenance of SCS integrity. The lack of robust compensation and synchronization mechanisms may lead to signal deterioration, increased error rates, and, as a result, a decline in the reliability and precision of SCSs.

4.3 Accountability Performance

The significance of accountability within the realm of satellite communications is paramount, particularly when evaluating the necessity for responsible and transparent governance of space resources by all involved parties, including nations, organizations, and private enterprises. In light of the swift expansion of

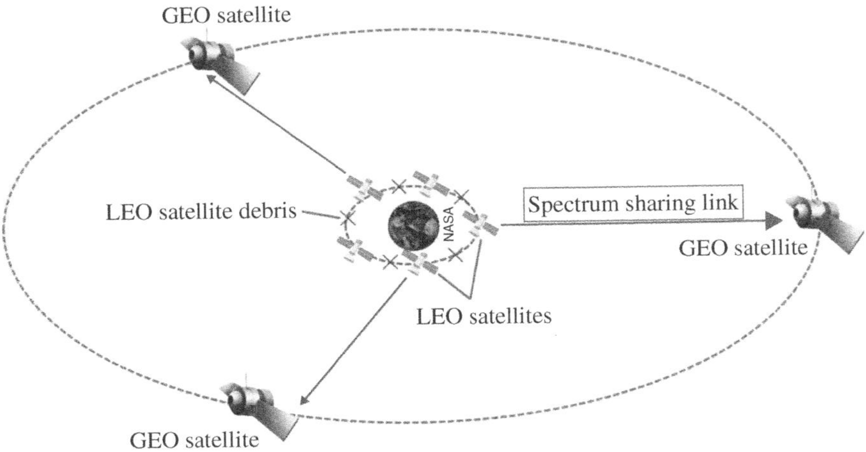

Figure 4.2 Accountability performance in multilayer satellites.

multilayer mega-constellations, accountability plays a crucial role in alleviating potential hazards, such as the creation of space debris, the risk of interference, and compliance with international regulatory standards.

1. **Debris Management and Responsibility:** As multilayer mega-constellations evolve, the proliferation of debris from transportation vehicles, along with both active and decommissioned satellites, exacerbates the escalating congestion in LEO. Accountability performance necessitates that satellite operators and spacefaring nations assume proactive stewardship over their space assets, guaranteeing compliance with sustainable practices such as debris remediation and the responsible disposal of defunct satellites. Operators are obligated to minimize the generation of space debris, implement effective collision avoidance strategies, and develop plans for deorbiting satellites once they cease to be operational. The ITU and other regulatory entities may establish frameworks for the monitoring and reporting of debris mitigation initiatives, thereby holding operators accountable for preserving the cleanliness and safety of the orbital environment [11].
2. **Interference Control and Accountability in Spectrum Sharing:** Due to the limited availability of spectral resources, SCSs are mandated to utilize the same frequency bands in accordance with ITU regulations. As depicted in Figure 4.2, accountability in this context pertains to the obligation of LEO operators to ensure that their systems do not generate excessive interference to GEO satellites, which are particularly vulnerable to signal attenuation owing to their stationary position in orbit. Operators are required to adopt interference mitigation strategies, including

power regulation, beam shaping, and frequency coordination, and must be accountable for adhering to the interference limits established by regulatory bodies. Furthermore, it is essential to maintain transparency in the management and allocation of spectral resources, with well-defined mechanisms for reporting any interference occurrences and resolving conflicts among operators.

3. **Operational Transparency and Logging:** SCSs, particularly those utilizing LEO mega-constellations, are required to meticulously document their operational activities. This documentation encompasses monitoring satellite positions, signal frequencies, and occurrences of interference or communication interruptions. Accountability performance mandates that these records are readily available to regulatory authorities and the public, thereby fostering transparency in satellite operations. In cases of interference or other operational anomalies, these records can prove to be essential resources for investigation and enforcement. Furthermore, by ensuring that satellite operators are entirely accountable for their actions, any deficiencies in regulatory compliance, such as failing to mitigate interference or effectively manage debris, can be traced back to the liable entity.
4. **Compliance with International Regulations:** Accountability encompasses the obligation to ensure adherence to the diverse legal and regulatory frameworks that oversee space operations. In this context, operators of LEO SCSs are required to align with the ITU's global coordination and regulatory standards to prevent interference among various satellite systems operating within identical frequency bands. Regulatory authorities, including the ITU, can enforce accountability by mandating satellite operators to provide evidence of their compliance with established spectrum-sharing protocols and debris mitigation strategies. Noncompliance with these standards may lead to penalties or limitations on the operator's capacity to launch or manage satellites.
5. **Liability and Financial Accountability:** With the escalating threats posed by space debris collisions and signal disruptions, accountability performance encompasses a dimension of financial responsibility. Satellite operators may be held liable for expenses related to debris removal, resolution of interference issues, and damages incurred by other satellites or terrestrial infrastructure. This responsibility could necessitate the establishment of insurance policies or financial guarantees to ensure that operators are equipped to manage potential costs resulting from their operational activities. Additionally, operators ought to be mandated to contribute to an international fund or insurance pool dedicated to debris removal and the preservation of the space environment.

6. **Enforcement Mechanisms:** To guarantee effective accountability in space operations, the implementation of robust enforcement mechanisms is imperative. These mechanisms may encompass real-time surveillance of satellite activities, independent assessments of space missions, and the establishment of international oversight organizations dedicated to monitoring satellite operations. In instances of violations, such as undue interference or negligence in debris mitigation, operators must incur quantifiable penalties or corrective measures. Such mechanisms ought to ensure that all participants in the satellite communication ecosystem are held accountable for their contributions to the sustainability of outer space.

   In summary, ensuring accountability within satellite communications, particularly regarding LEO mega-constellations, necessitates a holistic strategy to guarantee that satellite operators fulfill their responsibilities concerning the management of space resources, which encompasses debris mitigation, interference prevention, and adherence to regulatory standards. By establishing accountability among all relevant parties, the scholars can cultivate a more sustainable and collaborative space environment that not only facilitates the advancement of satellite communications but also reduces adverse effects on the space ecosystem.

4.4 Security-related Reliability Performance, Including Reliability, Latency, and Capacity/Throughput

4.4.1 Security-related Reliability Performance in SCSs

In the evolving landscape of satellite communication networks, the performance of network systems is increasingly scrutinized, particularly in the following aspects:

- Reliability, which refers to the ability of a network to function consistently without failure, is fundamental to guaranteeing continuous service.
- Latency, or the delay in transmitting data across the network, is a critical factor in applications requiring real-time communication, such as autonomous systems and interactive services.
- Capacity and throughput, which define the network's ability to handle large volumes of data, are essential for supporting bandwidth-intensive applications like video streaming and cloud computing.

The aforementioned key performance metrics are essential in ensuring that networks can adequately address the requirements of contemporary applications, particularly within the framework of advanced technologies such as 5G, 6G, and SCSs. Consequently, an in-depth comprehension of the interplay between these parameters and their influence on the overall performance of the network is imperative for the design and optimization of future communication infrastructures. This research investigates the relationship between security considerations and network reliability, analyzing how these elements collectively affect the overall efficiency and effectiveness of communication systems across diverse environments.

1. **Reliability:**
   Reliability constitutes a fundamental performance metric in SCSs, particularly within commercial and military domains where uninterrupted service is paramount. Reliability is typically quantified by the mean time between failures, which indicates the average duration of operation prior to a system encountering a failure. To bolster the reliability of SCSs, redundancy and fault tolerance strategies, such as supplementary satellites and multipath transmission, are frequently utilized. Furthermore, environmental factors, including radiation exposure and collision hazards, in conjunction with hardware malfunctions, can substantially influence the reliability of the system. Regarding security, encryption and authentication protocols serve effectively to thwart malicious attacks; however, they may also elevate system complexity and maintenance expenses. To mitigate these challenges, an integrated approach combining redundancy design with security protocols is crucial. Only by striking a balance between security and reliability requirements can SCSs guarantee efficient and stable operation. Fault tolerance and automatic recovery mechanisms are pivotal components in enhancing reliability, ensuring the system remains operational even amid single-point failures or external threats.
2. **Latency:**
   Latency serves as a pivotal performance indicator in SCSs, especially for real-time applications, including telemedicine, voice communication, and video conferencing. In the realm of satellite communication, latency is a result of both propagation delay and internal processing delay. The propagation delay is predominantly affected by the satellite's orbital classification. LEO satellites, owing to their closer proximity to Earth, generally provide reduced latency; conversely, GEO satellites encounter elevated propagation delays due to the increased distance from Earth. For example, an LEO satellite positioned at an altitude of 600 km experiences an uplink/downlink propagation delay of approximately 4 ms at the speed of light, which is typically negligible for applications such as voice calls.

   However, in addition to propagation delay, internal processing delays can also significantly contribute to heightened latency. These delays stem from the necessity for encryption, decryption, error correction, and data compression, which are vital for ensuring security and maintaining data integrity. For instance, the implementation of the G.729 speech codec can introduce an additional 10 ms processing delay during both the encoding and decoding phases [12]. Furthermore, channel coding and queuing delays associated with buffer management further exacerbate latency, potentially affecting the QoS in delay-sensitive applications such as telemedicine and emergency response.

In the realm of 6G networks, anticipated to facilitate ultra-reliable low-latency communication (URLLC) across a diverse array of applications, it is imperative to reduce latency while concurrently ensuring robust security measures. The total end-to-end latency experienced in SCSs comprises propagation delay, processing latency, and queueing delay. While encryption and various security protocols are vital for safeguarding data confidentiality and integrity, they can inadvertently contribute to increased latency. Consequently, to satisfy the rigorous latency demands of time-sensitive services, such as telemedicine and emergency response, satellite systems are required to implement lightweight encryption algorithms and enhance traffic management strategies, including traffic shaping and priority queueing, to guarantee that essential applications receive prioritized support without significantly jeopardizing security. To realize this equilibrium, it is critical to engineer low-latency security solutions that mitigate the influence of encryption and processing on latency, thereby ensuring an optimization of both security and performance. In doing so, SCSs can proficiently facilitate the secure transmission of real-time data, empowering applications such as telemedicine, voice communication, and emergency response to function effectively within a 6G framework.

3. **Capacity/Throughput:**
   Capacity and throughput serve as essential performance indicators for SCSs, directly affecting data transfer velocities and the number of users that the system can accommodate. Given that bandwidth is a finite asset in satellite communication, enhancements in system throughput and capacity necessitate the application of effective spectrum management and modulation methodologies. Techniques such as frequency-division multiplexing, time-division multiplexing, and BF enable satellite systems to facilitate simultaneous data transmission for multiple users utilizing the same frequency resources. The augmentation of capacity is also contingent upon the optimization of the system's architecture. For instance, multiple-input multiple-output (MIMO) technology, along with advanced coding and modulation techniques, significantly enhances bandwidth efficiency and bolsters data transfer capabilities. Concurrently, augmenting throughput necessitates an examination of the implications of security protocols on data transmission. Although encryption technologies are effective in safeguarding data confidentiality, they frequently introduce computational overhead that can adversely affect throughput. To achieve an equilibrium between throughput and security, SCSs may adopt

suitable encryption methodologies, such as employing symmetric encryption algorithms to mitigate computational delays and enhance throughput. Furthermore, traffic shaping and error correction technologies can be utilized to optimize data integrity during transmission while simultaneously increasing throughput.

4.4.2 Different Performance in LEO, MEO, and GEO Satellites

The reliability and security performance exhibit marked variations among LEO, MEO, and GEO systems, attributable to factors including orbital altitude, satellite longevity, and orbital inclination. LEO satellites, positioned in closer proximity to Earth, typically provide a reduced propagation delay; however, their frequent orbital adjustments and limited operational lifespans render them more vulnerable to failures, thereby complicating network management. To enhance the reliability of LEO satellites, low-latency security protocols are frequently integrated with redundancy frameworks and data recovery strategies. Conversely, GEO satellites, which occupy fixed orbital slots, benefit from extended lifespans and diminished satellite replacement frequencies, yet their increased distance from Earth leads to heightened signal propagation delays and greater susceptibility to space weather phenomena. Consequently, the reliability of GEO satellites is significantly reliant on ground station support and intricate satellite constellation oversight. MEO satellites, located between LEO and GEO, offer a compromise between propagation delay and reliability challenges, rendering them appropriate for medium-capacity and medium-latency applications. Throughout the three orbital classifications, the crux of optimizing satellite communication performance lies in the development of systems that not only guarantee reliability but also effectively mitigate security threats. Security measures such as encryption, authentication, and traffic monitoring are instrumental in thwarting external attacks, though they may adversely affect throughput and latency. Therefore, when architecting SCSs for each orbital category, it is imperative to strike a balance among security, reliability, latency, and throughput in accordance with the distinct application requirements.

Remarks: The QoS terminology includes numerous metrics, such as latency, traffic rate, BER, voice and video quality, etc.

Although all the aforementioned security and reliability criteria are critically significant, it poses considerable challenges to fulfill each of them concurrently. Indeed, it is often necessary to establish trade-offs. For instance, enhancing confidentiality in the uplink without employing robust secrecy coding typically entails diminishing the transmit power of the terminal. Nevertheless, this action concurrently results in a reduction of the system's integrity. Conversely, increasing the transmit power to bolster the integrity of the desired link inevitably escalates the risk of eavesdropping [13]. It is evident that further investigation is essential for the multicomponent Pareto optimization of the system to ascertain all optimal operating points within the nondominated set of solutions [14]. None of the metrics within these optimal solutions can be enhanced without compromising at least one or several of the other metrics.

References

1 Ki-Hong Park, Tian Wang, and Mohamed-Slim Alouini. On the jamming power allocation for secure amplify-and-forward relaying via cooperative jamming. *IEEE J. Sel. Areas Commun.*, 31(9):1741–1750, Sep. 2013.

2 Imre Csiszar and János Korner. Broadcast channels with confidential messages. *IEEE Trans. Inf. Theory*, 24(3):339–348, May 1978.

3 Matthieu Bloch, JoÃo Barros, Miguel Rodrigues, and Steven McLaughlin. Wireless information-theoretic security. *IEEE Trans. Inf. Theory*, 54(6):2515–2534, Jun. 2008.

4 Sabrina Gerbracht, Christian Scheunert, and Eduard Jorswieck. Secrecy outage in MISO systems with partial channel information. *IEEE Trans. Inf. Forensics Secur.*, 7(2):704–716, Apr. 2012.

5 Jun Xiong, Kai-Kit Wong, Dongtang Ma, and Jibo Wei. A closed-form power allocation for minimizing secrecy outage probability for MISO wiretap channels via masked beamforming. *IEEE Commun. Lett.*, 16(9):1496–1499, 2012.

6 Bo Bai, Wei Chen, and Zhigang Cao. Outage optimal subcarrier allocation for downlink secure OFDMA systems. In *2014 IEEE Globecom Workshops (GC Wkshps)*, pages 1320–1325, Dec. 2014.

7 Xiangyun Zhou, Matthew R McKay, Behrouz Maham, and Are Hjørungnes. Rethinking the secrecy outage formulation: A secure transmission design perspective. *IEEE Commun. Lett.*, 15(3):302–304, Mar. 2011.

8 Peishun Yan, Yulong Zou, Xiaojin Ding, and Jia Zhu. Energy-aware relay selection improves security-reliability tradeoff in energy harvesting cooperative cognitive radio systems. *IEEE Trans. Veh. Technol.*, 69(5):5115–5128, May 2020.

9 Yingbin Liang, Vincent Poor, and Shlomo Shamai. Secure communication over fading channels. *IEEE Trans. Inf. Theory*, 54(6):2470–2492, 2008.

10 Jiawei Wang, Chunxiao Jiang, and Linling Kuang. Turbo iterative DSSS acquisition in satellite high-mobility communications. *IEEE Trans. Veh. Technol.*, 70(12):12998–13009, Dec. 2021.

11 Pengwenlong Gu, Rui Li, Cunqing Hua, et al. Cooperative spectrum sharing in a co-existing LEO-GEO satellite system. In *Proc. IEEE Globecom Workshops*, pages 1–6, Taipei, Taiwan, 7–11 Dec. 2020.

12 Gerhard Schroder and Mostafa Hashem Sherif. The road to G.729: ITU 8-kb/s speech coding algorithm with wireline quality. *IEEE Commun. Mag.*, 35(9):48–54, Sep. 1997.

13 Yulong Zou, Jia Zhu, Xuelong Li, et al. Relay selection for wireless communications against eavesdropping: A security-reliability trade-off perspective. *IEEE Netw.*, 30(5):74–79, Sep. 2016.

14 Jingjing Wang, Chunxiao Jiang, Haijun Zhang, et al. Thirty years of machine learning: The road to pareto-optimal wireless networks. *IEEE Commun. Surv. Tutor.*, 22(3):1472–1514, Third Quart. 2020.

Part III

Security Enhancement Solutions

5

Active Security Enhancement Schemes

Active security enhancement schemes, including quantum communication, blockchain, THz, space-based laser communications, AI, and space situational awareness (SSA), aim to prevent or actively detect impending deleterious issues and support the stable operation of LEO satellites.

Quantum communication constitutes a symmetric secret key negotiation protocol capable of maintaining information-theoretic security and has evolved from academic research to off-the-shelf commercialization [1]. Furthermore, the blockchain is capable of satisfying the security requirement of decentralization, making LEO SCSs more robust. Furthermore, the progress of THz and laser-based communications is conducive to dealing with the CCI caused by the spectrum crunch. As a further advance, AI-assisted intelligent data-driven model-based solutions are suitable for traffic prediction, telemetry-based data mining, and anomaly detection. Finally, SSA employs many ground-based facilities or spaceborne facilities and novel algorithms to detect and track debris, thus cleaning up existing debris and substantially reducing the risk of collision.

5.1 Quantum Communication

5.1.1 Method Descriptions

Quantum communication signifies a transformative evolution in secure communication technologies, harnessing the fundamental principles of quantum mechanics to deliver unparalleled levels of security. The principal characteristic of quantum communication systems is the secure conveyance of information via quantum states, which exhibit heightened sensitivity to manipulation by eavesdroppers, thereby ensuring that any unauthorized interception of the communication is promptly identifiable.

Safeguarding Satellite Communications: Issues, Challenges, and Solutions, First Edition.
Jianping An, et al.

An essential component of quantum communication is QKD, a protocol that facilitates the sharing of a secret key between two parties over a potentially insecure channel, which can subsequently be utilized for encrypted communication. The most conceptually straightforward encryption technique involves the generation of a PR secret key, followed by the application of the modulo-two function to both the key and the information intended for encryption, referred to as plaintext. Naturally, the length of the key must be equivalent to that of the data sequence being transmitted, resulting in an overhead of 100%. The resulting encrypted information, termed ciphertext, may then be conveyed from the source to the destination via a public channel. With access to the secret key, the intended recipient can reconstruct the original plaintext utilizing the shared secret. Given that the key must remain confidential between the communicating parties, it necessitates secure transmission through an authenticated channel.

In contrast to classical cryptographic frameworks, which depend on the computational difficulty of specific problems to safeguard data, QKD is grounded in the fundamental laws of quantum mechanics to guarantee that the key remains impervious to eavesdropping. While straightforward principles may be employed in QKD systems for the processes of encryption and decryption, the negotiation of the secret key is contingent upon a quantum channel, in conjunction with an insecure public channel and an authenticated public channel. Should an eavesdropper endeavor to measure the quantum state of the key during its transmission, the state will collapse, thereby disclosing the presence of the intruder and permitting the legitimate parties to discard the compromised key and select an alternative.

5.1.2 Advantages and Disadvantages

The principal advantage of QKD resides in its theoretical invulnerability to hacking, as underpinned by the laws of quantum physics. Distinct from conventional encryption methods, which depend on the computational complexity of particular problems, the security of QKD is predicated on quantum principles, ensuring that any effort to eavesdrop on the communication is immediately detectable. This characteristic renders it a formidable solution against interception, thereby providing unprecedented confidentiality for sensitive information.

The medium employed for key negotiation in QKD is typically founded on optical fibers and free-space transmission. Optical fibers exhibit a minimal attenuation of approximately 0.3 dB/km and high stability, thereby rendering them more advantageous for the transmission of quantum signals.

QKD predominantly utilizes optical fibers or free-space pathways for transmission. Optical fibers are especially conducive for quantum signal propagation due to their low loss and high stability, thereby ensuring the integrity of communication over relatively short distances. Recent advancements in QKD technologies have resulted in enhancements in both the secret-key rate and the operational distance, thereby rendering QKD increasingly feasible for practical applications.

Despite its auspicious prospects, QKD continues to confront a multitude of unresolved challenges. A principal constraint of QKD pertains to its operational distance capabilities. Specifically, existing QKD networks predominantly facilitate point-to-point key distribution or short-distance network services, relying heavily on optical switches and routers. Explicitly, their operational range is restricted due to the necessity that quantum-domain signals remain unamplified; otherwise, they revert to the classical domain. Ongoing research and development in this domain are imperative to enable large-scale implementations. Furthermore, QKD fundamentally operates as a key negotiation and distribution protocol, wherein the secret key is utilized by classical systems. In contrast, quantum secure direct communication (QSDC) represents a comprehensive quantum communication protocol wherein confidential messages are transmitted directly over a quantum channel, obviating the need for a secret key, which has consequently facilitated its swift advancement.

In the context of free-space optical communication (FSO)-based QKD and QSDC scenarios, meteorological conditions such as cloud cover may adversely impact the transmission efficacy of both QKD and QSDC due to the dispersion caused by atmospheric turbulence. The presence of water molecules within the cloud layer engenders scattering and absorption of optical signals, while the attenuation of these signals by the cloud results in a diminishment of signal strength. Moreover, free-space QKD and QSDC typically necessitate the meticulous alignment of telescopes. Under cloudy conditions, the transmission of both visible and infrared light is severely restricted, thereby complicating the accurate alignment of telescopes. This misalignment adversely influences the stability of the QKD and QSDC systems. Compounding these issues, the high mobility of satellites may further aggravate the aforementioned challenges.

5.1.3 Applications

The family of satellite-based QKD systems was richly characterized in [2], along with various satellite channels using detailed examples.

In 2016, a significant landmark in the realm of satellite-based QKD was accomplished with the deployment of the world's inaugural quantum satellite, which successfully executed a QKD experiment in the space environment. This event heralded the inception of satellite-based quantum communication and unveiled novel prospects for secure communication on a global scale. Since that time, considerable advancements have been realized, with numerous nations and organizations, including European Space Agency (ESA) and Canada, striving to deploy their respective satellite-based QKD systems. These initiatives aim to establish a

Table 5.1 Major achievements in the implementation of QKD.

2017	1,200 km satellite-to-ground QKD at 1.1 kbit/s
2017	1,000 km satellite-to-ground entanglement-based QKD at 3.5 bit/s
2018	7,600 km apart ground gateways with satellite relay QKD and encryption demonstration at key volume 100 kB
2019	Continuous-variable QKD over 100 km fiber link at 0.14 kbit/s based on a photonic integrated quantum system
2020	Point-to-point discrete-variable QKD over 509 km fiber link at 0.1 bit/s
2020	First wavelength division multiplexing of 194 continuous-variable QKD at 172.6 Mbit/s over 25 km
2020	Point-to-point continuous-variable QKD over 13 km fiber link at 0.88 Mbit/s
2020	Point-to-point continuous-variable QKD over 202.81 km fiber link at 6.214 bit/s
2020	Discrete-variable QKD over 1,200 km free space optical link at 31 bit/s using Micius
2020	Continuous-variable QKD over 180 km fiber link at 31 bit/s based on a photonic integrated quantum system
2020	1,120 km apart ground stations entanglement-based QKD at 0.12 kbit/s
2021	4,600 km apart ground stations entanglement-based QKD at 47.8 kbit/s

global network for secure communication, capitalizing on the distinctive advantages of satellite-based infrastructure, which can surmount the distance limitations inherent in terrestrial optical fiber networks. The notable milestones attained in the implementation of QKD SCSs are systematically cataloged in Table 5.1.

The ongoing advancements in satellite-based QKD are propelling the evolution of a next-generation quantum communication framework capable of facilitating ultra-secure key distribution across extensive distances. By employing satellites to enable QKD, these systems can forge secure communication links between remote locations on Earth, thereby ensuring that sensitive data can be transmitted with unequivocal security, even in the presence of potential eavesdroppers. As an increasing number of countries allocate resources toward this technology, the aspiration for a comprehensive global quantum communication network becomes increasingly attainable, heralding a transformative epoch of secure global communication.

5.2 Blockchain

5.2.1 Method Descriptions

Blockchain is a decentralized, distributed digital ledger technology that stores transactional records across multiple databases, referred to as "blocks" [3]. These blocks are sequentially interconnected, thereby forming a "chain."

The fundamental tenet of blockchain is that each block encapsulates a record of transactions that are securely linked with the preceding block, culminating in an immutable and tamper-proof archival history of data. Transactions undergo verification through consensus mechanisms, wherein network participants authenticate the precision of the data prior to its incorporation into the blockchain. This architectural design obviates the necessity for a central authority, facilitating peer-to-peer networks that exhibit enhanced resilience to failures and security threats.

Within the framework of SCSs, blockchain technology can be employed to securely administer satellite communications and user data, thereby assuring that the information is decentralized and safeguarded against manipulation. Blockchain proffers a transparent and accountable paradigm that mandates all participants to validate and concur on the legitimacy of each transaction, rendering it resistant to fraudulent activities and cyber-assaults.

5.2.2 Advantages and Disadvantages

The blockchain methodology addresses several of the aforementioned security and reliability imperatives, specifically confidentiality, accountability, and decentralization. By depending on an encrypted database, users are required to possess the correct key to read information from this database or to write to it. Furthermore, once the information is updated, all pertinent data is collectively updated as a block and appended to the preceding version, thereby establishing an immutable and tamper-proof record. The premise is that the majority of participants scrutinize and authenticate this information; otherwise, the data cannot be modified on the blockchain.

Decentralization represents another salient attribute of the blockchain technique. In the event of a failure occurring within one or more nodes of a blockchain network, the remaining nodes maintain their data, and the network persists in its operational capacity. Consequently, a blockchain is frequently characterized as a distributed ledger, as the information resides across multiple devices within a peer-to-peer network, wherein each device replicates and retains an identical copy of the ledger and updates it autonomously, thereby providing high fault tolerance and resilience against attacks.

As a consequence of its distributed ledgers and consensus protocols, the blockchain methodology demonstrates immunity to DoS and DDoS attacks [4], as its decentralized framework negates the presence of a singular point of failure. This characteristic renders it particularly suited for securing satellite-based systems that are exposed to these categories of cyber threats.

Nonetheless, blockchain-based security services typically necessitate substantial computational power and storage capacity. For resource-constrained terminals and LEO satellites, this may present a considerable challenge, as the implementation of such systems might require significant innovation and optimization to mitigate overhead. As the quantity of nodes within a blockchain network amplifies, so too does the storage and computational encumbrance. Although blockchain offers robust security, its scalability for massive networks, such as those utilized in LEO satellites, can pose a limiting factor, necessitating efficient designs to preserve operational performance.

In blockchain implementations, user data is shared among a multitude of participants (such as satellites or gateways). Although blockchain technology guarantees data integrity, the exposure of users' sensitive information is a plausible concern, thereby leading to potential security vulnerabilities and issues pertaining to data misuse that may occur without the users' conscious awareness. Consequently, it is imperative that the safeguarding of user privacy is meticulously considered throughout the blockchain implementation process.

5.2.3 Applications

Given the escalating quantity of satellites and the corresponding user base supported by SCSs, the administration of their security presents a novel challenge. In response to this issue, the blockchain technique emerges as a promising paradigm for the secure and decentralized management of SCSs. The decentralized nature of blockchain guarantees that no single satellite or gateway possesses all critical information, thereby mitigating the likelihood of data breaches. Moreover, by facilitating robust encryption and consensus-based validation, blockchain effectively inhibits unauthorized alterations or tampering with transmitted data, thereby augmenting the overall security of SCSs.

Additionally, the fault tolerance and robustness of blockchain technology are especially beneficial for the management of high-mobility LEO satellites. Given that satellites are constantly moving and their communication links are susceptible to disruption, the decentralized and distributed characteristics of blockchain enable uninterrupted operation, even in scenarios where certain satellites or communication channels experience failure. This attribute positions blockchain as an optimal candidate for ensuring secure, reliable, and decentralized governance of LEO SCSs.

5.3 THz/Laser-based Spaceborne Communications

5.3.1 Method Descriptions

THz and laser-based spaceborne communication systems represent two nascent technologies that possess substantial promise for the advancement of high-velocity, high-capacity communication networks in space.

THz waves occupy the spectral region situated between microwave and infrared wavelengths within the electromagnetic spectrum, generally ranging from 0.1 to 10 THz. These frequency ranges facilitate ultra-high data transmission rates, enabling the conveyance of substantial volumes of information within condensed timeframes. THz communication applications extend to a variety of space-related functions, such as satellite communications, scientific instrumentation, and ISLs. The primary challenge associated with these technologies pertains to atmospheric attenuation, alongside the imperative for the development of highly efficient THz sources and detectors, with ongoing research aimed at creating compact, energy-conserving THz components suitable for practical applications in both terrestrial and space environments.

Laser communication leverages photonic emissions in the form of laser beams to convey data between spacecraft or from spacecraft to terrestrial ground stations. This modality offers a markedly higher bandwidth relative to conventional RF communication systems, thus facilitating expedited data transmission rates and more effective utilization of the available spectral resources. Laser communication systems function at optical wavelengths, typically within the near-infrared or visible light spectrum, which allows for precise targeting and beam steering—critical factors for effective long-range communication in space.

5.3.2 Current Developments

The frequency allocations of several commercial LEO satellite constellations are shown in Figure 5.1. It is evident that a considerable number of satellites function within the decimeter and centimeter wave frequency bands, as exemplified by satellites such as Iridium and Globalstar. The millimeter-wave (mmWave) frequency band is garnering scholarly interest due to its abundant spectral resources [5]. Numerous LEO satellite manufacturers, including Boeing, Starlink, and OneWeb, have petitioned for authorization to deploy satellites operating within the frequency range of 50.2–52.4 GHz [6]. Nevertheless, these frequency resources are experiencing increasing congestion. A plausible resolution involves elevating the operational frequency to the THz or even optical bands.

THz frequencies, typically exceeding 100 GHz, remain predominantly underutilized by ITU and possess the capacity to mitigate the congestion prevalent in lower frequency bands. Nevertheless, the presence of water vapor molecules and various propagation effects imposes constraints on the transmission distance within the THz band, rendering it unsuitable for satellite–Earth communication links [7]. Consequently, the application of THz technology is more appropriately aligned with ISLs situated above the Earth's atmosphere, where path loss is diminished and line-of-sight conditions exhibit greater favorability. THz transmitters and receivers may be engineered to circumvent the limitations associated with microwave frequency bands. Although the attenuation experienced

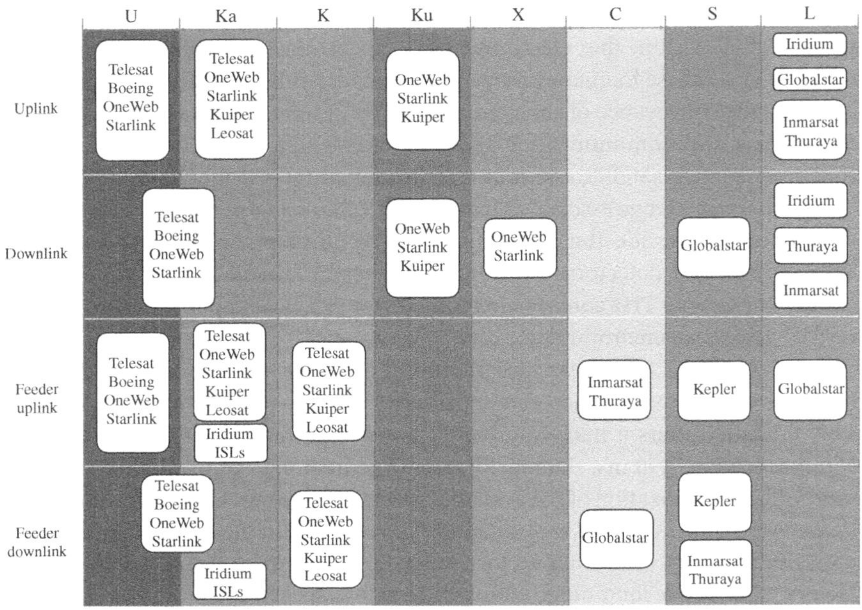

Figure 5.1 Frequency allocations of several commercial constellations between 1 and 60 GHz.

in the THz band is significant, this drawback may be potentially mitigated through the deployment of large-scale antennas utilized for BF on spaceborne payloads. The beamwidth of large-scale antennas operating within the THz band is narrower compared to that of conventional microwave ISLs, thereby enhancing their capacity to thwart eavesdropping attempts.

However, the greatest communication distance achieved was 21 km at 140 GHz, which is inadequate for ISLs. Consequently, the development of large antenna arrays and high-power devices functioning within the THz band is imperative to surmount the exceptionally high propagation losses and power constraints experienced by spaceborne transceivers in severe operational environments.

The laser band resides significantly above the electromagnetic spectrum. As a result, it exhibits a robust capacity for anti-interference. Laser communications remain undetectable by spectrum analyzers or RF meters due to the highly directional nature of the laser beam, thus rendering it a compelling candidate for ISLs and cross-layer communications. Furthermore, laser technology presents numerous advantages over microwave communications concerning dimensions, weight, and power dissipation when compared to the millimeter-wave band under equivalent data rate conditions.

5.3.3 Applications

Numerous research institutions globally have undertaken a plethora of experiments, which are concisely encapsulated in Table 5.2. Furthermore, Starlink evaluated the efficacy of "space lasers" for data transmission between two satellites, successfully relaying hundreds of gigabytes of information in September 2020. Starlink is actively engaged in the implementation of additional laser cross-links among their satellites to reduce the reliance on terrestrial infrastructure and to enhance connectivity in remote regions [8].

Table 5.2 The evolution of space-based laser communications.

Year	Project	Type	Country/ region	Data rate (Mbps)	Modulation	Distance (km)
2001	SILEX	GEO-LEO	Europe	50	IMDD	45,000
2006	OICETS	LEO-OGS	Japan	50	IMDD	610
2010	TerraSAR	LEO-OGS	Europe	5,625	BPSK	500–1,000
		LEO-LEO				1,000–5,000
2011	BTLS	LEO-OGS	Russia	125	IMDD	400
2013	LLCD	Lunar-OGS	US	622	PPM	400,000
2013	Alphasat	GEO-LEO	Europe	1,800	BPSK	45,000
2014	OPALS	LEO-OGS	US	50	IMDD	400
2014	SOTA	LEO-OGS	Japan	10	OOK/ IMDD	642
2016	MICIUS	LEO-OGS	China	5,120	DPSK	1,500
2016	OCSD	LEO-OGS	US	200	IMDD	450
2017	VSOTA	LEO-OGS	Japan	10	——	1,000
2017	SJ-13	GEO-OGS	China	4,800	IMDD	36,000
2020	EDRS-C	GEO-LEO	Europe	1,800	BPSK	45,000
2020	SJ-20	GEO-OGS	China	10,000	OOK/BPSK/ QPSK	36,000
2023	Cube-SOTA	GEO-LEO	Japan	10,000	DPSK	39,693
		LEO-OGS				1,103
2025	EDRS-D	GEO-GEO	Europe	3,600–10,000	BPSK	80,000
2025	ScyLight	GEO-LEO	Europe	100,000	——	——
		LEO-OGS				80,000

Since THz and laser beams exhibit a high degree of directionality, they encompass a restricted coverage area. This limitation presents formidable challenges for the alignment of signals (encompassing acquisition and tracking) within the domain of high-mobility LEO satellites. In contemporary satellite–Earth communication experiments, the studies outlined in Table 5.2 were executed between satellites and an optical ground station (OGS). The OGS utilizes its spatial positioning in conjunction with that of the satellites to facilitate signal alignment. Nevertheless, the mitigation of the Doppler shift necessitates the implementation of advanced processing algorithms to counteract its detrimental effects. Moreover, the alignment of signals proves to be more complex for ISLs within SCSs. The elevated velocity and jitter associated with spaceborne payloads render the precise alignment of the beam particularly challenging. Furthermore, even more pronounced Doppler frequency shifts may be detected for the spaceborne laser terminals in the "reverse seam," wherein adjacent satellites traverse in opposing directions.

The application of THz and laser communication within satellite communication stands to transform the field by delivering markedly enhanced data rates and optimizing the utilization of the available spectrum. However, their highly directional characteristics and vulnerability to Doppler frequency shifts present substantial obstacles, especially within LEO SCSs. As scholarly inquiry persists in addressing these impediments, including the development of more sophisticated signal processing methodologies and adaptive tracking systems, THz and laser communication are likely to emerge as pivotal facilitators for next-generation satellite communication frameworks, thereby supporting high-bandwidth applications and global coverage.

5.4 AI-based Solutions

In recent years, the advancements in AI within terrestrial wireless communication frameworks have been successfully extrapolated to SCSs. The deployment of AI encompasses a multitude of functionalities, including data classification, anomaly detection, and predictive analytics, which possess the capability to adjust in accordance with real-time operational fluctuations inherent in space missions.

The integration of AI within satellite systems confers significant benefits, including enhanced decision-making processes, automation of security protocols, and the capacity to identify and address potential threats promptly. Through the assimilation of empirical data, AI can adaptively respond to emerging patterns and challenges, a necessity for safeguarding satellite communications against the continuously evolving landscape of threats.

Among the various AI methodologies, ML has developed and established itself as an instrumental apparatus adept at learning empirical models and monitoring alterations in data patterns throughout space missions [9]. ML engages in the training of its model utilizing historical datasets, modifies its parameters accordingly, and subsequently applies the acquired patterns to novel, previously unencountered data for purposes such as classification, regression, or anomaly detection. This characteristic of ML is particularly beneficial in the realm of security solutions as it facilitates the examination of diverse data types from multiple perspectives. This section will elucidate how AI can be employed to process and analyze various forms of satellite data, which include tele-traffic data, housekeeping data, power system data, and spectrum data.

5.4.1 Method Descriptions and Classifications

5.4.1.1 Tele-traffic Data

Tele-traffic data encompasses information pertinent to communication traffic, which includes both the magnitude and the categories of data transmitted among satellites, as well as between satellites and ground stations. Within the framework of SCSs, tele-traffic data constitutes a pivotal element as it facilitates the management and optimization of the deployment of communication resources, including but not limited to bandwidth, transmission power, and routing pathways. The proficient administration of this data is indispensable for ensuring the delivery of high-quality services, averting congestion, and enhancing the overall efficacy of SCSs.

ML models are exceptionally adept at scrutinizing tele-traffic data, owing to their capacity to process extensive datasets, discern patterns, and generate predictions grounded in historical data. A principal application of ML in the realm of tele-traffic data resides in the anticipation of future traffic patterns. Through the training of models utilizing historical traffic datasets, ML algorithms possess the capability to project traffic loads, thereby facilitating improved resource allocation, such as the dynamic scheduling of communication resources. Beyond traffic forecasting, ML is also applicable for the identification of anomalies within tele-traffic data. Unusual traffic patterns, which frequently signify nefarious activities such as cyberattacks or network intrusions, can be recognized by training ML models on established traffic behaviors. Upon the identification of such abnormal patterns, the system is positioned to implement proactive measures to avert congestion and security breaches prior to their escalation.

The precise processing of tele-traffic within SCSs is of utmost importance. Forecasted traffic patterns are integral to the optimization of routing pathways and the prescheduling of networking resources, thereby alleviating CCI and reducing both transmission outages and the likelihood of congestion. Furthermore, the early identification of abnormal traffic instigated by malevolent entities can preemptively thwart congestion from manifesting. The extreme learning machine (ELM) represents a neural network characterized by a straightforward architecture and rapid computational speed, rendering it adept at accurately delineating the traffic load of LEO satellites [10]. Additionally, an advanced long short-term memory (LSTM) prediction model founded on a deep recurrent neural network (RNN) has been proposed [11], which serves as an exemplary ML methodology for managing multivariate time-series data, given its capability to model the intricate interactions within a system. Nevertheless, the LSTM prediction model entails a considerable computational load. To mitigate this, a gated recurrent unit (GRU)-based neural traffic prediction algorithm, featuring a streamlined gate structure, has been developed [12], whose training efficiency and accuracy can be significantly enhanced through the integration of robust techniques such as transfer learning and online training.

5.4.1.2 Housekeeping Data

The housekeeping data acquired from the satellites comprises a plethora of measurements and readings that epitomize the operational status of the satellite as well as its surrounding atmospheric conditions. Through the examination of irregularities within the housekeeping data, one can deduce potential system failures and imminent alerts, thereby enabling the satellite to undertake preemptive measures aimed at mitigating the risk of operational failure. Housekeeping data characteristically exhibits high dimensionality, multimodality, and heterogeneity [13], rendering it a multifaceted and invaluable resource for ML applications.

The implementation of AI has witnessed a significant increase for the purposes of processing and analyzing housekeeping data, attributed to its capacity to discern intricate patterns and anomalies that may herald potential system failures. A prevalent application is anomaly detection, wherein AI algorithms discern atypical behaviors or outliers within the dataset that could indicate malfunctions or operational risks. LSTM networks have been employed for the task of anomaly detection, as they proficiently process time-series data while delivering predictions regarding the prospective state of the system [14]. Furthermore, AI models leverage probabilistic clustering methodologies to scrutinize multimodal data, thereby facilitating the identification of complex interrelationships among various satellite subsystems and external variables.

In the realm of SCSs, the continuous monitoring and analytical assessment of housekeeping data is imperative for ensuring system reliability and preempting potential failures. By employing ML techniques, satellite systems can evaluate historical housekeeping data to forecast future behaviors and recognize early indicators of failure. The temporal convolution network has also garnered considerable attention for time-series forecasting, exhibiting favorable parallel processing capabilities and temporal characterization [15], which demonstrates enhanced operational efficiency when juxtaposed with LSTM. The projected telemetry data derived from the aforementioned predictive techniques can subsequently undergo further analysis to identify any prospective failures.

A conventional methodology involves juxtaposing the prediction error with a predetermined or dynamically adjusted threshold to ascertain which data points are anomalous [16].

5.4.1.3 Power System Data

The data pertaining to power systems is of paramount importance in ensuring the secure and stable functioning of a designated mission. Nevertheless, the architecture of the power system is intricate, wherein faults may manifest in components such as cables, solar arrays, batteries, power distribution switches, and power controllers, among others. Evaluating the correlation among various fault types and establishing the causative links between these faults and their associated predictive indicators present a significant challenge.

Fortunately, a substantial volume of power system data can be collected, thereby enabling ML methodologies to develop a precise fault classification model that delineates the relationship between anomalous data and fault associations [17]. Specifically, the LSTM technique elucidated earlier has the capacity to forecast the operational efficacy of solar arrays, which are recognized for exhibiting the highest failure rates among satellite power components, by scrutinizing historical performance data alongside environmental variables [18]. Through the continuous surveillance of power system data, ML models are poised to identify early indications of deterioration or failure, such as power fluctuations or irregularities in battery performance, and subsequently initiate preemptive measures.

5.4.1.4 Spectrum Data

Spectrum data exhibits temporal dependencies and is frequently influenced by a multitude of factors, including interference from other communication systems, ambient environmental noise, and the dynamics of satellite movement. Consequently, the continuous monitoring of spectrum data and the identification of potential issues, such as signal degradation or interference, is imperative for ensuring the stability and operational efficiency of SCSs.

ML tools play a pivotal role in the processing of spectrum data within SCSs to detect anomalies and interference. Analogous to other forms of time-dependent telemetry data, spectrum data can be utilized within an LSTM-based predictive model to facilitate the intelligent management of the future signal spectrum, as well as to identify anomalies and interference [19]. Moreover, features associated with interference can be employed for the classification of interference to identify and mitigate sources of interference [20].

5.4.2 Future Directions of AI

The predominant methodologies in ML for executing prediction, anomaly detection, and classification tasks on satellite data predominantly include neural networks. A notable instance of such methodologies is LSTM, which represents a

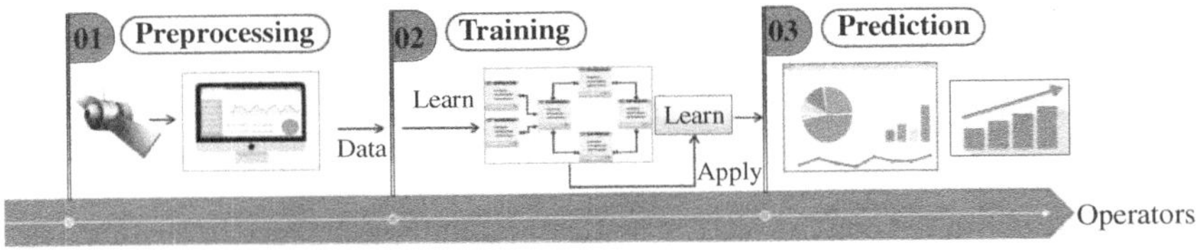

Figure 5.2 The entire process of the ML-based security solutions for satellite data.

specific category of neural network characterized by recurrent connections. These recurrent connections facilitate the retention of significant information within the network's internal memory. The GRU bears similarities to LSTM in its incorporation of recurrent connections and memory retention; however, it is distinguished by a more streamlined architecture and a reduced number of information pathways compared to LSTM. Both LSTM and GRU are classified as specialized variants of conventional RNNs, wherein each neuron acquires its input from the preceding time step and transmits the output to the subsequent time step. In contrast, ELM is categorized as a feedforward neural network that lacks recurrent connections. In essence, ELM does not possess explicit memory to archive historical information; however, it is renowned for its rapid training capabilities. Figure 5.2 delineates the comprehensive process of the ML-based security solutions conceptualized for satellite data as examined in this chapter [21].

AI-driven security mechanisms for satellite communications can derive significant advantages from these ML methodologies by improving their capacity to foresee prospective security infringements, identify irregularities, and categorize network behaviors with greater precision. Nonetheless, the amalgamation of AI within satellite frameworks must also contend with the challenge of achieving an equilibrium between performance efficiency and resource limitations, especially in light of the restricted computational and energy resources inherent to satellites. Continued investigation into lightweight AI architectures and their effective deployment within satellite networks will be crucial for actualizing their comprehensive potential in fortifying satellite-terrestrial communications.

5.5 Space Situational Awareness

5.5.1 Method Descriptions

The proliferation of resident space objects (RSOs), encompassing satellites, spacecraft, and space debris, in Earth's orbital vicinity has escalated significantly, thereby presenting a formidable threat to activities conducted in outer space. In response to this urgent challenge, governmental bodies, military organizations, and space agencies have instituted SSA programs aimed at providing collision alerts and facilitating debris mitigation [22]. These SSA initiatives leverage an extensive array of sensors, including terrestrial radar systems, optical telescopes, and spaceborne radar (SBR) technologies, to deduce the orbital characteristics of objects with a focus on their classification and identification. The overarching objective of SSA is to yield accurate data pertaining to the positioning, trajectory, and attributes of RSOs, thereby diminishing the likelihood of collisions and ensuring the protection of space operations [23].

An auspicious approach to SSA entails the amalgamation of SBR and spaceborne cameras (SBCs) for the purposes of detecting and monitoring debris [24]. Spaceborne technologies, exemplified by radar prototypes functioning at elevated frequencies (e.g., 94 GHz for debris of centimeter dimensions or 340 GHz for THz radar), have been engineered to produce high-resolution three-dimensional representations of space debris. Specifically, the inverse synthetic aperture radar (ISAR) methodology has been utilized to augment the imaging fidelity of rotating debris. Additionally, coordinated formations of satellites equipped with cameras or radars can collaboratively ascertain the three-dimensional position of debris, thereby enabling precise tracking.

Moreover, another critical facet of SSA involves the application of advanced algorithms to monitor debris. Both filtering algorithms [25] and Bernoulli filtering [26] methodologies are employed to enhance the accuracy of the estimations concerning the state and position of the debris, ensuring that multiple sensors can synergistically contribute to a more precise depiction of the debris trajectory. Furthermore, space-based sensors utilize Bayesian inference techniques to improve trajectory forecasting and refine debris detection capabilities over protracted durations.

5.5.2 Advantages and Disadvantages

Ground-based radar systems and optical telescopes are exceptionally adept at facilitating the observation of RSOs. Nevertheless, both ground-based radar systems and optical telescopes exhibit their respective advantages and disadvantages. Ground-based radar systems possess the capability to operate continuously irrespective of atmospheric conditions; however, one notable drawback pertains to their substantial operational costs arising from the necessity for high transmitter power. Moreover, ground-based radar systems are unable to precisely detect debris in LEO scenarios with a diameter less than 10 cm [27], attributable to their considerable aperture dimensions. Conversely, optical telescopes offer significant sensitivity for observation, yet their functionality is constrained by varying weather conditions, which limit the duration of their observational capabilities [23].

In response to these challenges, the concept of utilizing SBR or SBC for debris detection and tracking has been proposed, as these systems are positioned closer to space debris, thereby diminishing the necessity for high-power transmission and enhancing their efficiency in identifying and monitoring smaller debris.

The efficacy of spaceborne solutions that employ THz radar technology necessitates further enhancement. Spaceborne systems present the potential to protect information transmitted between LEO satellites through THz-based ISLs. Additionally, these systems are capable of detecting space debris, thereby safeguarding the satellite through the utilization of THz-based radar [27]. However, the maximum achievable transmission power of spaceborne THz apparatus imposes significant constraints on both communication range and radar detection distance, thereby necessitating the development of large-scale space-borne antenna arrays and high-power devices as a focal point for future research endeavors.

References

1 Yuan Cao, Yongli Zhao, Qin Wang, et al. The evolution of quantum key distribution networks: On the road to the Qinternet. *IEEE Commun. Surv. Tutor.*, 24(2):839–894, Second Quart. 2022.
2 Nedasadat Hosseinidehaj, Zunaira Babar, Robert Malaney, et al. Satellite-based continuous-variable quantum communications: State-of-the-art and a predictive outlook. *IEEE Commun. Surv. Tutor.*, 21(1):881–919, First Quart. 2019.
3 Zhou Zhou, Youliang Tian, Jinbo Xiong, et al. Blockchain-enabled secure and trusted federated data sharing in IIoT. *IEEE Trans. Industr. Inform.*, 19(5):6669–6681, May 2023.
4 Yuntao Wang, Zhou Su, Jianbing Ni, et al. Blockchain-empowered space-air-ground integrated networks: Opportunities, challenges, and solutions. *IEEE Commun. Surv. Tutor.*, 24(1):160–209, First Quart. 2022.
5 Ibrahim A Hemadeh, Katla Satyanarayana, Mohammed El-Hajjar, et al. Millimeter-wave communications: Physical channel models, design considerations, antenna constructions, and link-budget. *IEEE Commun. Surv. Tutor.*, 20(2):870–913, Second Quart. 2018.
6 Shiyi Xia, Quanjiang Jiang, Cheng Zou, and Guotong Li. Beam coverage comparison of LEO satellite systems based on user diversification. *IEEE Access*, 7:181656–181667, Dec. 2019.
7 Qing Xia and Josep M Jornet. Multi-hop relaying distribution strategies for terahertz-band communication networks: A cross-layer analysis. *IEEE Trans. Wirel. Commun.*, 21(7):5075–5089, Jul. 2022.
8 Quan Chen, Giovanni Giambene, Lei Yang, et al. Analysis of inter-satellite link paths for LEO mega-constellation networks. *IEEE Trans. Veh. Technol.*, 70(3):2743–2755, Mar. 2021.
9 Zhenping Li. A machine learning solution for satellite health and safety monitoring. *J. Space Oper. Commun.*, 18(1):1–24, Jan. 2022.
10 Zhenyu Na, Zheng Pan, Xin Liu, et al. Distributed routing strategy based on machine learning for LEO satellite network. *Wirel. Commun. Mob. Comput.*, 2018:3026405, Jul. 2018.
11 Chen Han, Aijun Liu, Liangyu Huo, et al. A prediction-based resource matching scheme for rentable LEO satellite communication network. *IEEE Commun. Lett.*, 24(2):414–417, Nov. 2019.

12 Ning Li, Lang Hu, Zhongliang Deng, et al. Research on GRU neural network satellite traffic prediction based on transfer learning. *Wirel. Pers. Commun.*, 118(1):815–827, Jan. 2021.

13 Takehisa Yairi, Naoya Takeishi, Tetsuo Oda, et al. A data-driven health monitoring method for satellite housekeeping data based on probabilistic clustering and dimensionality reduction. *IEEE Trans. Aerosp. Electron. Syst.*, 53(3):1384–1401, Jun. 2017.

14 Yakun Wang, Jianglei Gong, Jie Zhang, et al. A deep learning anomaly detection framework for satellite telemetry with fake anomalies. *Int. J. Aerosp. Eng.*, 2022:1–9, Jan. 2022.

15 Yuan Wang, Yan Wu, Qiong Yang, and Jun Zhang. Anomaly detection of spacecraft telemetry data using temporal convolution network. In *2021 IEEE International Instrumentation and Measurement Technology Conference (I2MTC)*, pages 1–5, Glasgow, United Kingdom, 17–20 May 2021. IEEE.

16 Kyle Hundman, Valentino Constantinou, Christopher Laporte, et al. Detecting spacecraft anomalies using LSTMs and nonparametric dynamic thresholding. In *Proceedings of the 24th ACM SIGKDD International Conference on Knowledge Discovery & Data Mining*, pages 387–395, Jul. 2018.

17 Hui Li, Jing He, Xiaowei Wang, and Hui Yang. Research review and prospect of fault diagnosis method of satellite power system based on machine learning. *DEStech Trans. Comput. Sci. Eng.*, 332:543–549, Mar. 2019.

18 Jingyi Dong, Yuntong Ma, and Datong Liu. Deep learning based multiple sensors monitoring and abnormal discovery for satellite power system. In *2019 International Conference on Sensing, Diagnostics, Prognostics, and Control (SDPC)*, pages 638–643, Beijing, China, 15–17 Aug. 2019. IEEE.

19 Lachlan Gunn, Peter Smet, Edward Arbon, et al. Anomaly detection in satellite communications systems using LSTM networks. In *2018 Military Communications and Information Systems Conference (MilCIS)*, pages 1–6, Canberra, ACT, Australia, 13–15 Nov. 2018. IEEE.

20 Jiangyi Qin, Fei Zhang, Kai Wang, et al. Interference signal feature extraction and pattern classification algorithm based on deep learning. *Electronics*, 11(14):2251, Jul. 2022.

21 Ijaz Ahmad, Jani Suomalainen, Pawani Porambage, Andrei Gurtov, Jyrki Huusko, and Marko Höyhtyä. Security of satellite-terrestrial communications: Challenges and potential solutions. *IEEE Access*, 10:96038–96052, 2022.

22 Bin Li, Jian Huang, Yanming Feng, et al. A machine learning-based approach for improved orbit predictions of LEO space debris with sparse tracking data from a single station. *IEEE Trans. Aerosp. Electron. Syst.*, 56(6):4253–4268, Dec. 2020.

23 Davide Cataldo, Luca Gentile, Selenia Ghio, et al. Multibistatic radar for space surveillance and tracking. *IEEE Aerosp. Electron. Syst. Mag.*, 35(8):14–30, Aug. 2020.

24 Marco Maffei, Augusto Aubry, Antonio De Maio, et al. Spaceborne radar sensor architecture for debris detection and tracking. *IEEE Trans. Geosci. Remote Sens.*, 59(8):6621–6636, Aug. 2021.

25 Roberto Opromolla and Alessia Nocerino. Uncooperative spacecraft relative navigation with LIDAR-based unscented Kalman filter. *IEEE Access*, 7:180012–180026, Dec. 2019.

26 Xinglin Shen, Zhiyong Song, Hongqi Fan, et al. General Bernoulli filter for arbitrary clutter and target measurement processes. *IEEE Signal Process. Lett.*, 25(10):1525–1529, Oct. 2018.

27 Xu Yang, Yiming Pi, Tong Liu, et al. Three-dimensional imaging of space debris with space-based terahertz radar. *IEEE Sens. J.*, 18(3):1063–1072, Feb. 2018.

6

Passive Security Enhancement Schemes

6.1 Advanced Security-oriented Antennas

6.1.1 Method Descriptions

In the contemporary digital landscape, characterized by the critical importance of information security, sophisticated security-centric antenna technologies have emerged as essential for guaranteeing both the integrity and confidentiality of communications. A primary strategy employed in this domain is eavesdropping mitigation, which seeks to obstruct unauthorized entities from intercepting communication signals and gaining access to sensitive information, thus preserving the privacy of the data being transmitted. Conversely, jamming mitigation is concerned with neutralizing malevolent jamming signals to maintain the uninterrupted and stable functionality of communication links. To effectively confront the myriad security threats that arise during communication processes, it is essential to conduct a comprehensive analysis of these countermeasure strategies. The subsequent sections offer an in-depth elucidation of these methodologies, with a particular focus on the mitigation of eavesdropping and jamming.

6.1.1.1 Eavesdropping Mitigation

Recent investigations have increasingly concentrated on sophisticated security-centric antennas designed for secure communications, owing to their capability to augment the radiation pattern directed toward the designated receiver while attenuating emissions in the majority of alternative directions. Nevertheless, a potential eavesdropper equipped with a highly sensitive receiver may still succeed in intercepting the communication link via a side lobe. To counteract this susceptibility, side-lobe randomization methodologies can be implemented to alleviate information leakage through the side lobes. Moreover, these advanced antennas employ beamforming (BF) and artificial noise (AN) in the downlink, strategically

Safeguarding Satellite Communications: Issues, Challenges, and Solutions, First Edition.
Jianping An, et al.

directing AN toward prospective eavesdroppers to proactively obstruct interception endeavors [1–3].

However, it is plausible that eavesdroppers may capture the primary lobe signal from various locations situated between a LEO satellite and the terrestrial surface. This situation is notably common in LEO SCSs, as satellites perpetually traverse above, consequently positioning potential eavesdroppers within the coverage area of the primary lobe. Under such circumstances, the traditional PA demonstrates limitations due to its beam patterns being exclusively dependent on the angle. Conversely, the implementation of the frequency diverse array (FDA) can be utilized to address this constraint [4].

In [5], the authors presented a linear frequency diverse array (LFDA) that produces a beam pattern that is contingent upon both the angle and distance by systematically shifting the carrier frequencies across its antenna array. Nevertheless, due to the inherent coupling between the beam pattern's length and direction, it remains feasible for eavesdroppers to intercept the communications of legitimate users from specific vantage points. To ameliorate this predicament, the research documented in [6–8] put forth various nonlinear frequency offset methodologies, including logarithmic, exponential, and random offsets, with the objective of decoupling the spatial dimensions of distance and direction within the beam pattern. Figure 6.1 delineates the beam patterns exhibited by the conventional PA, LFDA, and logarithmic FDA. While the peak position of the PA is invariant with respect to distance, the peak positions of both the LFDA and logarithmic FDA are influenced by the parameters of direction and distance. Importantly, the peaks generated by the LFDA manifest an "S" shape and display periodic characteristics, with the periodicity being dictated by the frequency offset employed. In contrast, the logarithmic FDA effectively decouples distance from the angle, resulting in the formation of a focused spot beam. Comprehensive derivations pertaining to these phenomena can be located in [5, 6].

6.1.1.2 Jamming Mitigation

In the contemporary multifaceted and ever-evolving realm of communications, sophisticated security-focused antennas provide exceptional performance benefits. A particularly significant characteristic of these antennas is their capacity to meticulously modify the beam direction pattern in reaction to fluctuating power interference scenarios. This functionality is critical for maintaining both the integrity and security of communication systems [9].

For an antenna array with N elements, the radiation pattern $P(\theta, \phi)$ can be described as

$$P(\theta, \phi) = \left| \sum_{n=1}^{N} w_n e^{jk\mathbf{r}_n \cdot \hat{\mathbf{r}}} \right|^2 \tag{6.1}$$

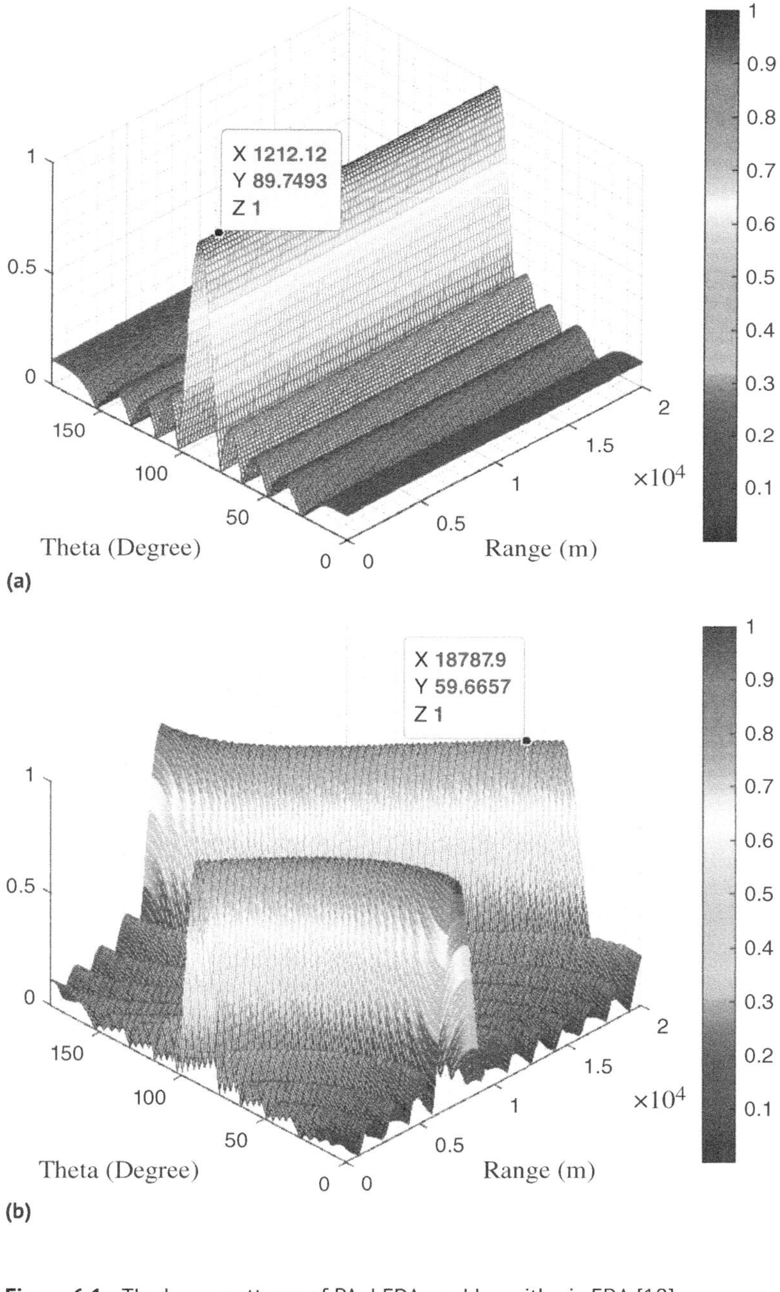

Figure 6.1 The beam patterns of PA, LFDA, and logarithmic FDA [10].

(Continued)

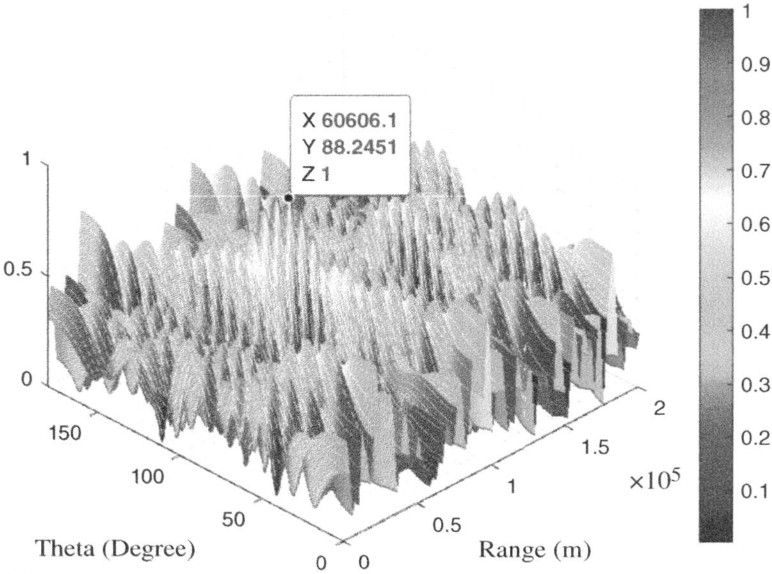

(c)

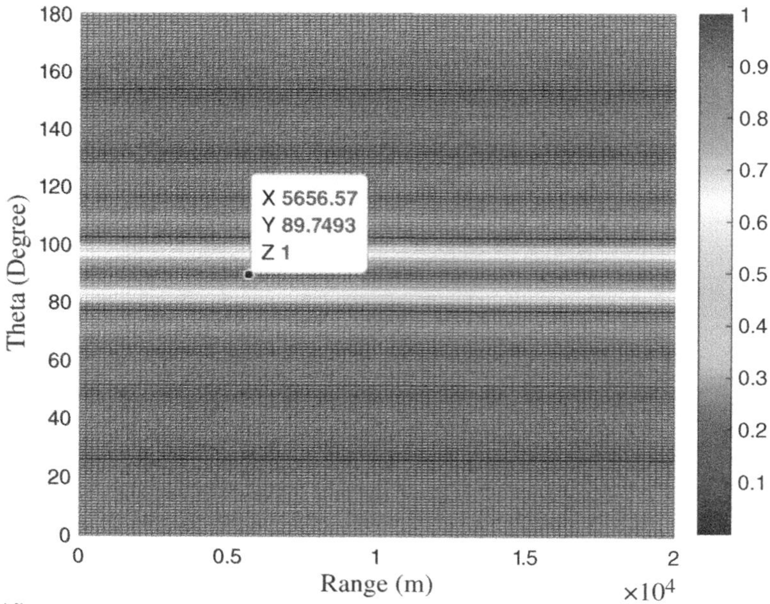

(d)

Figure 6.1 (Cont'd)

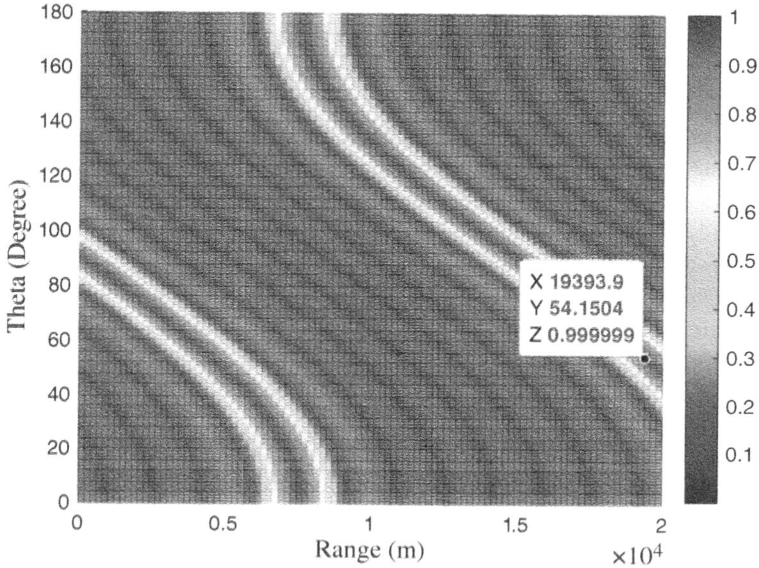

(e)

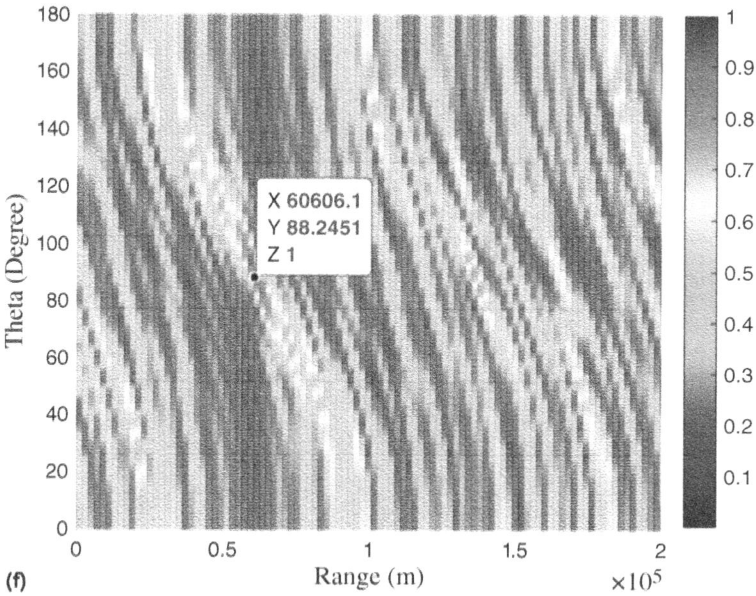

(f)

Figure 6.1 (Cont'd)

where w_n represents the weight (amplitude and phase) for the nth element, and $\mathbf{r}_n$ is the position vector of that element. By choosing appropriate weights, the antenna can direct the signal in the desired direction and suppress interference from others.

Specifically, as illustrated in Figure 6.2, these sophisticated antennas possess the capability to meticulously modify the beam direction pattern within the azimuthal angle. In practical communication contexts, sources of interference may emanate from a multitude of directions. Upon the detection of interference signals originating from either the left or right flanks of the antenna, the antenna system is able to swiftly initiate a response by employing intelligent algorithms to compute the optimal adjustment angle. Subsequently, it modifies the beam direction pattern in the azimuthal angle to effectively circumvent the interference source, thereby minimizing the adverse effects of interference from the left or right side of the antenna [11].

BF constitutes the principal methodology employed by antennas to mitigate interference and optimize signal reception. The fundamental tenet of BF encompasses the modulation of both the amplitude and phase of the signals at each individual antenna element within an array, thereby facilitating constructive interference in the targeted direction while inducing destructive interference in non-preferred directions.

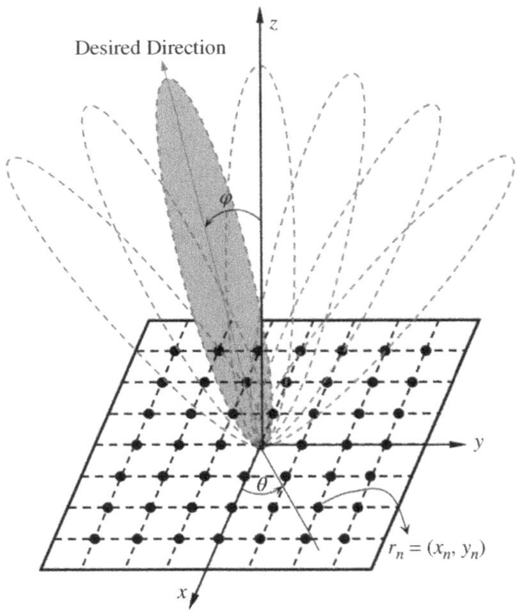

Figure 6.2 Control of beam direction with BF.

6.1.2 Applications

6.1.2.1 Eavesdropping Mitigation

In the domain of satellite communication, eavesdropping mitigation technologies assume a pivotal function in safeguarding the confidentiality and integrity of transmitted signals. These technologies are meticulously engineered to thwart unauthorized interception through the strategic regulation of signal propagation and the reduction of vulnerabilities inherent in communication links. Eavesdropping mitigation technologies possess significant applications:

- In LEO SCS, the unique orbital positioning of LEO satellites significantly elevates the risk of eavesdropping. The incorporation of FDA technology offers a robust solution to mitigate this concern. Through the generation of a beam pattern that is contingent upon both angular and distance parameters, the FDA enhances the precision of signal transmission management while simultaneously diminishing the likelihood of signal interception by eavesdroppers.
- In safeguarding the communication link between satellites and ground stations, sophisticated security-focused antennas assume a pivotal function. Augmenting radiation in the specified orientation while attenuating emissions in other directions guarantees the robust and secure conveyance of communication signals between satellites and ground stations, thereby significantly mitigating the potential for eavesdroppers to intercept signals emanating from side lobes or other angles.
- In the intricate scenario where eavesdroppers might be situated within the primary lobe direction in satellite communication, FDA technology employing nonlinear frequency offset methodologies, such as the spot beam generated by the logarithmic FDA, is capable of achieving a decoupling of distance and directionality. This advancement offers a more robust security assurance for satellite communication in multifaceted environments and enhances the overall security and stability of SCSs.

6.1.2.2 Jamming Mitigation

At present, the technology of multibeam antennas has experienced extensive implementation within the domain of satellite communication. The former paradigm of employing a singular beam to encompass a vast expanse of the Earth's surface has been rendered obsolete, supplanted by the utilization of multiple narrow beams to facilitate uninterrupted coverage across Earth or designated regions. Within satellite environments, sources of interference have the potential to emanate from a multitude of directions. For instance, communication signals relayed from satellites positioned in higher orbits to terrestrial locations or other LEO satellites may disrupt the signal reception capabilities of LEO satellites.

In such instances, sophisticated interference-mitigating antennas possess the capability to swiftly identify interference originating from the elevation angle and adapt the beam pattern accordingly to counteract the interference. This strategic adjustment ensures that the pathway for signal transmission circumvents the interference zone, thereby significantly alleviating the adverse effects of elevation-angle interference on the communication system [12].

6.1.3 Advantages and Disadvantages

6.1.3.1 Eavesdropping Mitigation

In the field of communication security, eavesdropping mitigation offers several advantages:

- Security-oriented antennas optimize radiation patterns to focus energy on the receiver, reducing eavesdropping risks.
- BF and AN in the downlink enhance resistance by directing beams and interfering with the reception of eavesdroppers.
- FDA technology generates beam patterns based on both angle and distance, improving security with nonlinear frequency offsets.

Despite significant progress, eavesdropping mitigation has some limitations:

- Side-lobe randomization cannot fully eliminate the risks of eavesdropping, especially from high-sensitivity receivers.
- In LEO satellite communication, eavesdroppers can intercept signals in the main lobe direction, exposing PA technology's limitations.
- The time-varying beam patterns of FDA technology limit its use in scenarios that require stable signals.

6.1.3.2 Jamming Mitigation

In the contemporary intricate landscape of communication, interference profoundly undermines the quality and stability of communication systems, whereby malicious signals precipitate data loss and inaccuracies. The technology designed for jamming mitigation is integral in preserving dependable communication links by effectively counteracting a multitude of interference sources. Numerous advancements have been made in the realm of jamming mitigation.

- Security-oriented antennas adjust beam direction to avoid interference, enhancing communication stability.
- BF technology suppresses interference from undesired directions while directing signals to the target receiver.
- Advanced antenna technology flexibly adjusts the direction of the beam in response to interference from multiple directions.

Nonetheless, no technological system is devoid of imperfections. Although jamming mitigation occupies a significant position, there exist several substantial limitations that cannot be overlooked. It is only through a thorough comprehension of these drawbacks that we can more effectively evaluate the advantages and disadvantages in practical implementations and facilitate the advancement of communication technology.

- High technical complexity due to the need for intelligent algorithms and complex BF.
- High consumption of computing resources to detect interference and calculate optimal adjustments.
- Reliance on accurate interference detection; errors can lead to ineffective adjustments and signal disruption.

6.2 Reconfigurable Intelligent Surface

In contemporary times, propelled by significant progressions in metamaterials and microelectromechanical systems, the reconfigurable intelligent surface (RIS) has surfaced as a potentially transformative technology within the domain of wireless communication.

6.2.1 Method Descriptions

The RIS disseminates a substantial quantity of passive microreflectors across a modifiable substrate and executes sophisticated control via an interconnected controller. The RIS possesses the capability to modify characteristics including the phase and amplitude of the reflectors in real time, facilitated by its advanced functionalities in areas such as adaptive modification, intelligent control algorithms, ML, and optimization, in addition to network integration and management, thereby attaining intelligent regulation and enhancement of the reflected signals. The technology underpinning RIS can be utilized in a myriad of communication frameworks and wireless networks, including the augmentation of signal coverage, the enhancement of system throughput, and the fortification of passive security measures.

As depicted in Figure 6.3, the implementation of a RIS in the context of enhancing secure communication facilitates the dynamic modulation of both the amplitude and phase of its reflective components, thereby introducing an element of diversity and stochasticity into the signal reflection pathway. This procedural adjustment guarantees that the reflected signals traversing various pathways remain statistically independent, notwithstanding potential correlations between legitimate and eavesdropping channels. Fundamentally, the RIS operates as a

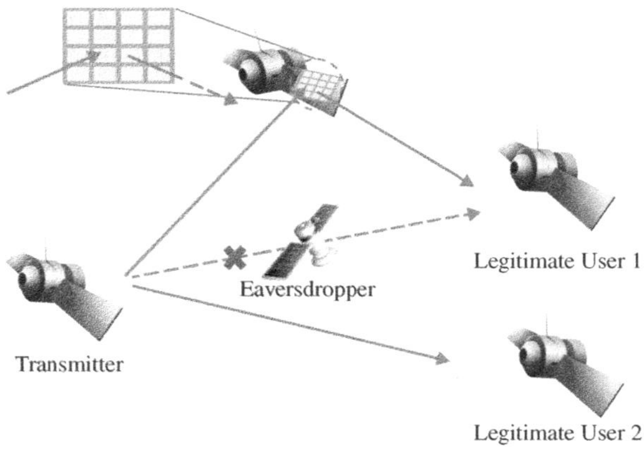

Figure 6.3 RIS in satellite passive security communication.

"secure link" for users, effectively obstructing eavesdroppers from acquiring pertinent information through the analysis of signal correlations.

Moreover, through the implementation of joint BF that integrates active transmission from the base station alongside passive reflection facilitated by the RIS, direct and reflected signals can be constructively synthesized for the legitimate user while simultaneously inducing destructive interference for potential eavesdroppers, thereby considerably augmenting the security performance. Additionally, given that the RIS operates as a passive entity devoid of active RF transmission components, it presents several advantages, including being lightweight, compact, adaptable, and economically viable. These characteristics render it particularly advantageous for extensive deployment within wireless networks, thereby guaranteeing secure communication with minimal energy expenditure and reduced implementation costs.

6.2.2 Applications

In the domain of satellite-based secure communication facilitated by RIS, numerous applications exist.

On one hand, it possesses the potential to facilitate cooperative jamming, thereby contributing to the enhancement of communication security. For instance, the RIS is capable of reflecting terrestrial jamming signals directed toward ground-based eavesdroppers endeavoring to intercept satellite downlink transmissions. Through the collaborative design of the transmit beamformer weights utilized by the terrestrial base station and the reflection coefficients of the RIS, the signal-to-interference-plus-noise ratio (SINR) experienced by eavesdroppers can be diminished within the interference tolerance constraints of satellite users, consequently fortifying security measures. Furthermore, an increase in the number of reflecting elements can augment the interference power aimed at eavesdroppers. Moreover, the RIS can be deployed via a HAP to safeguard the communication link established between a LEO satellite and a UAV receiver. Even in scenarios where CSI pertaining to eavesdroppers is absent, legitimate users can bolster system security in adverse environments by amplifying the number of reflecting elements. When either statistical or perfect CSI is accessible, optimization methodologies for the RIS can be devised to sustain an elevated level of security. Simultaneously, it has been observed that 3-bit quantization can effectively mitigate substantial degradation in secrecy performance.

On the other hand, active RIS finds application in the domain of satellite communication security. Figure 6.4 depicts the comprehensive spectrum of scenarios wherein intelligent reflective surfaces are employed to enhance passive security measures. A collaborative jamming strategy is implemented to aid the terrestrial network in protecting legitimate satellite downlink transmissions. Nevertheless, the influence of active RIS on the performance of secrecy has not been comprehensively elucidated. Empirical studies indicate that, under certain conditions, a meticulously designed active RIS exhibits superior secrecy energy efficiency compared to its passive counterpart. Nonetheless, the potential of active RIS to bolster security remains an unresolved inquiry. Regarding the protection of LEO satellite

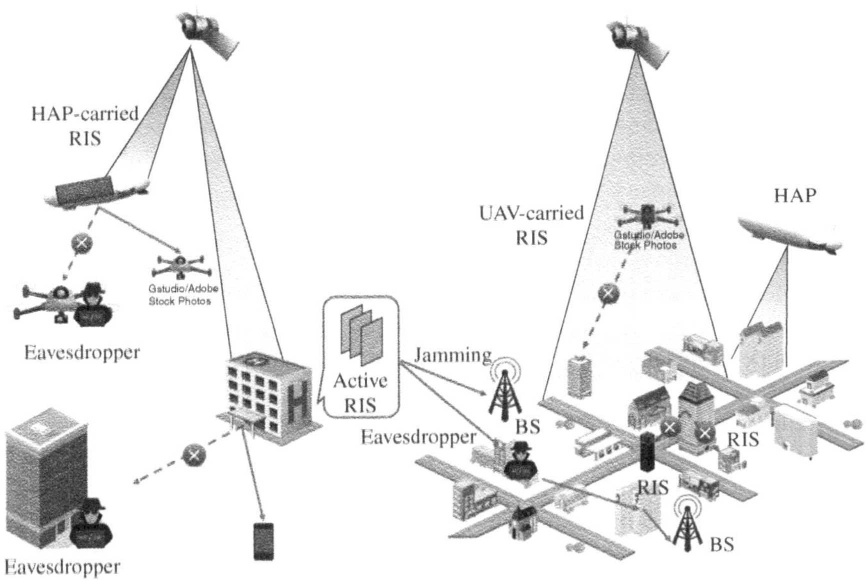

Figure 6.4 Application scenarios for RIS-enabled passive security enhancement solutions.

systems, a comparative analysis of the benefits associated with active versus passive RIS, as well as the implications of the amplification power budget, necessitates further scholarly exploration.

Furthermore, RIS is employed to alleviate system interference and CCI. In LEO satellite systems, the total data rate can be enhanced. Within the integrated ground-LEO satellite network that utilizes spectrum sharing, an increase in the quantity of RIS, the number of reflective elements, and the resolution of phase shifts results in a marked improvement in the received SINR. Additionally, RIS presents benefits in scenarios assisted by HAPs; however, challenges such as channel estimation inaccuracies and multipath propagation effects must be meticulously resolved during the design process. The deployment of RIS on UAVs can enhance signal propagation within terrestrial networks, mitigate interference produced by uplink signals transmitted from ground stations to satellites, and subsequently improve SINR for target ground users.

6.2.3 Advantages and Disadvantages

RIS, as a burgeoning technology in the realm of intelligent reflective systems, exhibits considerable promise in the augmentation of passive security measures. By virtue of its capacity to manipulate wireless channels without the need for the active transmission of supplementary signals, it can enhance the security of communication systems. Nonetheless, in spite of its merits, the deployment of RIS within LEO SCSs encounters a myriad of challenges, primarily attributable to the inherently dynamic characteristics of LEO trajectories and the intricate nature of the extraterrestrial environment. The ensuing discussion delineates the principal advantages and possible limitations associated with the utilization of RIS in the enhancement of passive security.

Advantages

1. **Enhancing Cooperative Jamming Effectiveness:** RIS possesses the capability to adeptly redirect terrestrial jamming signals toward potential eavesdroppers, consequently diminishing their SINR and significantly bolstering the security of communication. Furthermore, augmenting the quantity of reflective elements can further intensify the jamming effect, thereby rendering unauthorized eavesdropping increasingly arduous.
2. **Enhancing Security in Spectrum-sharing Frameworks:** The implementation of intelligent BF in conjunction with RIS has the potential to alleviate CCI while simultaneously enhancing the signal integrity for designated users. This functionality is especially advantageous within the context of integrated LEO satellite and terrestrial networks, thereby augmenting the overall security and dependability of the system.

3. **Energy Efficiency and Deployment Flexibility:** In contrast to active jamming methodologies, RIS functions by passively reflecting electromagnetic signals without necessitating supplementary power amplification, thereby rendering it a more energy-efficient alternative. Furthermore, it possesses the capability to be deployed with flexibility across terrestrial infrastructures, HAPs, or UAVs to suitably address a diverse array of application contexts.

Disadvantages

1. **Challenges Posed by the Elevated Mobility of LEO Satellites:** The swift transit of LEO satellites engenders highly variable channel conditions, thereby requiring the real-time optimization of RIS configurations. Neglecting this imperative may result in suboptimal efficacy in diminishing the SINR experienced by potential eavesdroppers, ultimately undermining the enhancement of security measures.
2. **Restricted Accessibility and Ecological Limitations:** Ground-based or UAV-mounted RIS can solely offer enhancements in security during limited intervals when LEO satellites are within LoS. Furthermore, operations of UAV-based RIS are restricted by energy constraints and environmental influences, including unfavorable meteorological phenomena, thereby diminishing their long-term utility.
3. **Engineering Challenges in LEO Deployment:** The direct implementation of RIS on LEO satellites, while simultaneously guaranteeing the augmentation of security, would result in an escalation of both satellite mass and launch expenditures. Moreover, such systems must be resilient to the severe conditions prevalent in the space environment, encompassing extreme fluctuations in temperature and exposure to cosmic radiation, both of which could adversely affect their reliability and operational longevity.

6.3 Spread Spectrum Techniques

For more than seven decades, SS technology has played a pivotal role in military communication systems. The augmentation of signal bandwidth enhances both confidentiality and resistance to jamming attempts. Established methodologies such as DSSS, FHSS, and multi-carrier DSSS (MC-DSSS) serve to protect information security, as potential interceptors face significant challenges in detecting confidential data without appropriating frequency-hopping (FH) patterns or codebooks.

6.3.1 Method Descriptions

DSSS technology has been extensively implemented in SCSs owing to its remarkable efficacy, particularly in safeguarding communications from unauthorized interception. The PSD of DSSS signals is generally characterized by a low magnitude. As illustrated in Figure 6.5, when these signals arrive at the receiver, they are frequently obscured by ambient noise, thereby rendering it exceedingly challenging for potential eavesdroppers to retrieve pertinent information.

Furthermore, DSSS technology demonstrates a significant level of resistance to jamming. Should a jamming signal endeavor to disrupt the authentic transmission, the receiver's correlator will disperse the jamming signal across the complete bandwidth throughout the despreading procedure, attributable to the intrinsic dissimilarity in the correlation between the jamming signal and the local pseudo-noise sequence. Conversely, the legitimate signal can be precisely despread back to its original, narrower bandwidth. This operation considerably enhances the SNR of the baseband data, owing to the inherent processing gain (PG). Consequently, the PSD of the jamming signal remains minimal in the baseband, thereby further bolstering the anti-jamming efficacy of DSSS.

It is important to note that the anti-jamming efficacy of DSSS technology is intrinsically linked to the PG. Nonetheless, the payload capacity of the system is limited by the ratio of bandwidth to the spreading factor, thereby underscoring a critical trade-off between transmission rate and anti-jamming efficacy within LEO SCSs. Specifically, in scenarios characterized by substantial interference, the extension of the DSSS spreading sequence length significantly bolsters the system's anti-jamming capabilities through an enhancement of the PG. However, this enhancement is accompanied by a concomitant decline in system throughput. In contrast, the pursuit of a heightened transmission rate necessitates the

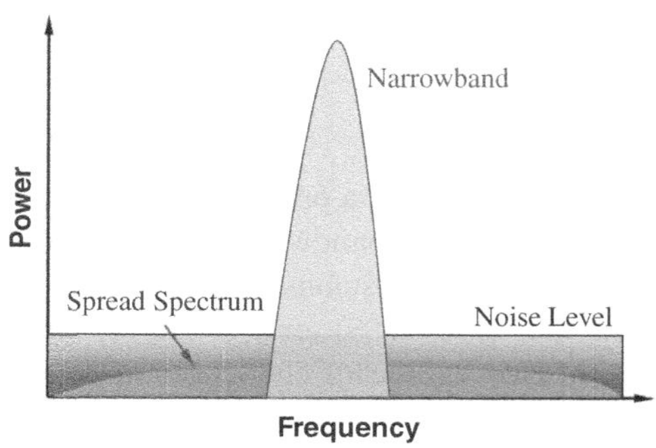

Figure 6.5 Principle of DSSS.

utilization of a shorter spreading sequence, which may inadvertently undermine the system's resilience to interference.

FHSS, a widely recognized anti-jamming technique, exhibits notable differences when compared to DSSS. Throughout the transmission phase, the FHSS transceiver dynamically transitions between various subcarrier frequencies in accordance with the SS code. Consequently, the bandwidth of the FHSS signal is composed of disparate frequency bands. To enhance the anti-jamming efficacy, FHSS technology is frequently integrated with cognitive radio technology. By employing an adaptive FH mechanism, it adeptly circumvents frequency bands that are adversely impacted by interference, thereby safeguarding the stability and dependability of communication.

The notion of fast frequency-hopping spread spectrum (FFHSS) arises from the comprehension of swift FH within a singular symbol. It is distinguished by a dwell time allocated for each hop that is inferior to the symbol duration, thereby facilitating the completion of multiple hops within the confines of a single symbol duration, consequently bestowing the system with a formidable anti-jamming capability. FFHSS is capable of employing low-complexity noncoherent de-hopping and demodulation techniques. Although this strategy may mitigate the complexities associated with system implementation, it unavoidably results in a considerable diminution of the SNR. In contrast, the coherent reception methodology of FFHSS demonstrates superior performance. Nonetheless, its implementation complexity also escalates markedly, thereby imposing more rigorous demands on system hardware and algorithmic processes.

In contrast to DSSS, the MC-DSSS receiver utilizes a more sophisticated spectrum sensing methodology. It possesses the capability to continuously monitor and analyze the currently operational frequency in real time with high accuracy, enabling it to effectively discern both available and occupied frequency bands, subsequently allowing for the dynamic formulation of subcarrier allocation strategies based on this analysis. On one hand, the MC-DSSS receiver has the option to circumvent existing signals by assigning each subcarrier to an available frequency band, thereby enhancing the integrity of the communication process. However, this approach renders the signals susceptible to interception by malicious entities, consequently leading to a deterioration in confidentiality. Conversely, the MC-DSSS receiver may also proactively obscure the subcarriers within the existing signals to bolster confidentiality. In this scenario, the operational frequency of each subcarrier is configured to align with that of the existing signals. Nonetheless, this tactic is likely to result in the legitimate signals being compromised by the existing signals, thereby undeniably diminishing the integrity of communication.

As illustrated in Figure 6.6, the waveform corresponding to MC-DSSS employing eight subcarriers is depicted. Within this waveform, a total of four subcarriers are designated to the available frequency bands, while the remaining four are

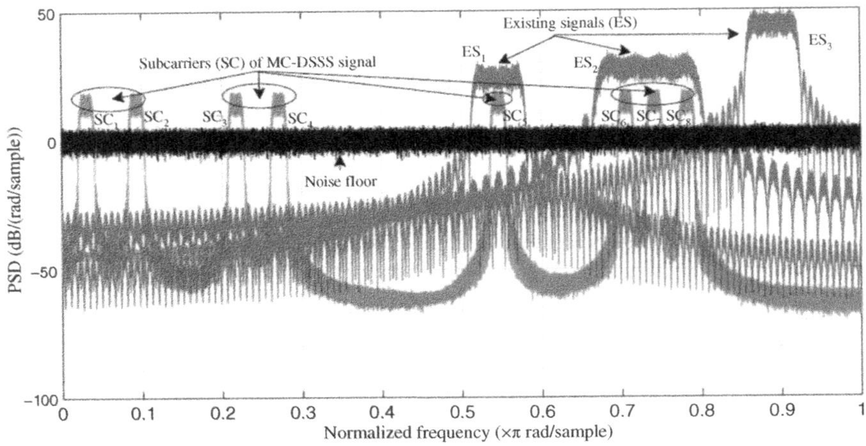

Figure 6.6 The trade-off between confidentiality and integrity in MC-DSSS systems [10].

strategically concealed within the current signals. The SNR for each subcarrier is quantified at 10 dB, whereas the interference-to-signal ratio (ISR) for the three preexisting signals in conjunction with each subcarrier interference signal is measured at 18 dB, 38.5 dB, and 25 dB, respectively. These parameters serve to elucidate the performance characteristics and challenges inherent to the MC-DSSS system within empirical applications.

6.3.2 Applications

Satellite communication encounters a multifaceted electromagnetic milieu, characterized by both natural and anthropogenic interferences, in addition to the potential threat of signal interception. SS technology, distinguished by its inherent advantages, has emerged as a critical mechanism for safeguarding the secure transmission of satellites.

6.3.2.1 Satellite IoT System

The satellite IoT system integrates IoT technology with satellite communication technology to facilitate the global interconnection of IoT devices. Chirp SS technology is employed for the multiuser uplink transmission within the satellite IoT framework. Signals originating from diverse user devices can be transmitted within the same frequency band subsequent to a spreading process, and at the receiving terminal, the information pertaining to each user can be precisely recovered through the processes of despreading and correlation. For instance, in the context of environmental monitoring in remote regions, the environmental data acquired by sensors strategically distributed across various locations is relayed back to the data center via the satellite IoT network. The SS technology guarantees the dependable transmission and privacy protection of substantial volumes of sensor data.

6.3.2.2 GPS Satellite Navigation System

GPS exemplifies a prominent application of satellite communication, wherein SS technology is of paramount significance. The signals emitted by GPS satellites utilize the code division multiple access (CDMA) SS technique, with each satellite employing a distinct pseudo-random noise (PRN) code for the modulation of its signal. Within an urban milieu characterized by numerous towering edifices, the signal is susceptible to multifaceted interference stemming from reflection and obstruction phenomena. The GPS receiver is capable of deriving pertinent information from intricate signal patterns through the execution of a correlation analysis utilizing the specific PRN code disseminated by the satellite.

6.3.2.3 Military Satellite Communication

Military satellite communication encompasses exceedingly stringent security and reliability prerequisites, with technology serving as a fundamental component in the assurance of communication security. In the context of military confrontations, adversarial entities may employ electronic countermeasures to disrupt and capture satellite communications. Illustrating this, the SCS utilized by the United States military integrates DSSS and FHSS methodologies. The DSSS technique broadens the signal across an expanded frequency spectrum, thereby attenuating the signal PSD and complicating detection and interference efforts by adversaries. Concurrently, the FHSS approach facilitates rapid transitions of the signal among varied frequencies, effectively circumventing interference-prone frequency bands in accordance with a predetermined FH sequence. Within military operations, numerous distinct units necessitate simultaneous communication with satellites. The multiple-access functionality inherent in SS technology allows various users to utilize the same frequency band through the application of diverse SS codes, thereby achieving non-interfering communication.

6.3.3 Advantages and Disadvantages

SS communication, serving as a significant technological instrument within the contemporary communication domain, possesses distinct advantages while simultaneously presenting certain substantial limitations that warrant consideration.

Advantages

1. **High Confidentiality:** SS communication augments the signal bandwidth, facilitating its amalgamation with background noise, thereby ensuring data protection from eavesdropping. For instance, military communications implement SS methodologies for the secure conveyance of classified information.
2. **Excellent Anti-jamming Ability:** Various SS techniques deliver formidable anti-jamming capabilities. DSSS distributes interference across the bandwidth, thereby enhancing the SNR. FHSS employs FH to circumvent interference, thereby ensuring the stability of communication links. FFHSS executes rapid frequency transitions, thereby bolstering resistance to intricate interference patterns.
3. **High Flexibility:** MC-DSSS employs spectrum sensing to assign subcarriers predicated on real-time frequency analysis. This system can be modified to circumvent interference or enhance confidentiality, contingent upon the specific communication context.

Disadvantages

1. **Limited Transmission Efficiency:** In DSSS, an augmentation in the spreading factor to mitigate jamming effects concurrently diminishes the data throughput. This inherent trade-off, particularly evident in LEO SCSs, impedes transmission velocities, as illustrated in satellite-ground station interactions where interference necessitates extended DSSS sequences.
2. **SNR Loss:** The utilization of a noncoherent demodulation technique in FFHSS simplifies system architecture; however, this simplification concomitantly leads to a reduction in SNR, thereby adversely impacting the quality of communication. Conversely, the implementation of coherent reception enhances performance metrics but simultaneously escalates system complexity and associated expenditures.
3. **Difficulty in Balancing Confidentiality and Integrity:** Within the framework of MC-DSSS, the emphasis on confidentiality through the obfuscation of subcarriers undermines transmission integrity, whereas a focus on maintaining integrity may jeopardize confidentiality, thereby compromising data precision, particularly in time-sensitive applications such as video streaming.

6.4 Non-SS Jamming Suppression Techniques

In the realm of satellite communication, the transmission of signals frequently encounters a multitude of interference-related challenges. When the power of the interference surpasses the maximum tolerance threshold of the SS receiver, traditional anti-interference strategies often become ineffective. Consequently, the SS system is necessitated to implement specialized interference mitigation algorithms, including time-domain adaptive filtering, which is capable of processing signals within the time domain, as well as transform-domain adaptive filtering, which adeptly identifies and manages interference within the frequency domain, thereby ensuring the stability and reliability of the communication process.

6.4.1 Method Descriptions

Temporal-domain adaptive filtering algorithms are particularly effective in mitigating narrowband interference. The least mean square (LMS) algorithm is frequently selected as a design alternative owing to its computational simplicity [13, 14]. The fundamental principle underlying the LMS algorithm is to replicate a causal Wiener filter by iteratively adjusting the filter coefficients until the LMS of the error signal is minimized. This methodology employs stochastic gradient descent, indicating that the filter coefficients are modified solely based on the error at the present symbol instant. In the case of a conventional LMS algorithm, the rate of convergence is influenced by the step-size parameter (μ), which can be systematically decreased as the process approaches convergence to the optimal minimum.

On one hand, an increase in the value of μ correlates with an accelerated convergence of the weights. Consequently, we are able to swiftly monitor and address the oscillating jamming. On the other hand, an elevated μ results in an increased variance of the weights, which subsequently impacts the efficacy of jamming mitigation. Hence, the effective implementation of the LMS algorithm necessitates a careful balancing act.

In contrast, transform-domain adaptive filtering possesses the capability to swiftly monitor narrow-band jamming variations without necessitating an iterative methodology [15, 16]. The transform-domain adaptive filtering algorithm represents a sophisticated advancement in signal processing technology, primarily employed in contexts that involve the management of interfering signals, such as the attenuation of narrow-band jamming within SS systems. This algorithm conducts its operations on the received signal within the frequency domain. Initially, it discerns the interfering components present in the signal, accurately identifying the specific source of interference amid a complex amalgamation of information. Thereafter, through the application of band-pass filtering techniques, it mitigates or modifies the interfering elements. Upon the completion of these essential processing stages within the frequency domain, the signal undergoes a transformation back into the time domain, thereby facilitating the enhanced utilization of the processed signal by the system. Furthermore, one of the algorithm's noteworthy benefits is its ability to promptly adapt to the variations of narrow-band jamming without reliance on an iterative process, thereby significantly enhancing signal quality and overall system performance.

6.4.2 Applications

6.4.2.1 Time-domain Adaptive Filtering in Satellite Secure Communication

Interference Suppression: Within the domain of satellite communication, a multitude of interference sources exist. For instance, terrestrial electromagnetic interference and co-frequency interference arising from other SCSs exemplify this phenomenon. In instances where these interferences manifest as relatively stable narrow-band disruptions, the application of time-domain adaptive filtering can reveal its inherent advantages. Utilizing the LMS algorithm as a representative case, it is capable of incrementally modifying the filter coefficients to attenuate interference in accordance with the statistical properties of the interference signal.

It is posited that a satellite acquires communication signals within a predetermined frequency band emanating from terrestrial sources; however, it is also subjected to a relatively stable narrow-band interference source in proximity, characterized by minimal fluctuations in both frequency and power over a specified duration. By employing a time-domain adaptive filtering algorithm, the incoming signal can be effectively processed within the time domain. The adaptive filter is designed to continuously modify its coefficients in response to the discrepancy between the received signal and the target signal, which represents the original signal in the absence of interference. As time progresses, the algorithm facilitates a systematic attenuation of the interference signal's influence, thereby enhancing the capability of the receiving system to accurately demodulate the pertinent communication signal, ultimately safeguarding the integrity and dependability of satellite communication.

Signal Tracking: In contexts where the attributes of the signal exhibit gradual variations, the implementation of time-domain adaptive filtering proves effective in accurately tracking the signal. Signals transmitted via satellite communication may experience gradual alterations attributable to variables such as propagation distance and atmospheric conditions.

For example, when the communication link established between a satellite and a terrestrial station induces alterations in the signal's delay, amplitude, and other characteristics that gradually evolve due to atmospheric refraction, the time-domain adaptive filter possesses the capability to modulate the step-size parameter appropriately in order to effectively monitor such variations in the signal. It updates the weights in accordance with the instantaneous error associated with each symbol, thereby adapting to gradual fluctuations in the signal, which ensures the integrity of the received signal, mitigates signal distortion, and preserves the seamless functionality of satellite-based communication.

6.4.2.2 Transform-domain Adaptive Filtering in Satellite Secure Communication

Interference Suppression: Within the intricate electromagnetic milieu of satellite communications, the frequency and amplitude of interference signals exhibit rapid variations. Transform-domain adaptive filtering possesses the capability to promptly adjust to the oscillations of narrow-band jamming under such circumstances.

For instance, in scenarios where a satellite encounters multiple narrow-band interference signals emanating from various directions, and the frequencies of such interference signals exhibit the propensity to fluctuate unpredictably, the merits of transform-domain adaptive filtering in the analysis of received signals within the frequency domain become markedly evident. Initially, it transposes the received signal into the frequency domain, where it can distinctly delineate the frequency bands that contain the interference components. Subsequently, through methodologies such as band-pass filtering, it effectively mitigates these interference signals with precision. Upon reverting the processed signal to the time domain, it substantially alleviates the adverse effects of interference on the satellite communication signal, thereby safeguarding communication integrity.

Spectrum Resource Management: The allocation of spectrum resources for satellite communication is of paramount importance. The implementation of transform-domain adaptive filtering can significantly enhance the efficacy of spectrum management. Through the processing of signals within the frequency domain, one can attain a more precise comprehension of spectrum occupancy.

For instance, in instances where it is imperative to ascertain the presence of unauthorized interference signals encroaching upon the spectrum allocated for satellite communication, transform-domain adaptive filtering techniques can facilitate an analysis of the received signal within the frequency domain. The prompt identification and mitigation of such interference signals that unlawfully occupy the spectrum not only safeguard the integrity of satellite communication but also enhance the efficient allocation of spectrum resources and augment the overall spectral efficiency of the SCS.

6.4.3 Advantages and Disadvantages

In the domain of signal processing, adaptive filtering within the temporal and transform frameworks constitute significant methodologies for addressing perturbative signals. Each technique possesses distinct characteristics and serves divergent functions in applied contexts. The ensuing discussion aims to elucidate their respective merits and demerits.

6.4.3.1 Time-domain Adaptive Filtering

Advantages

1. **Reduced Algorithmic Complexity:** Consider the LMS algorithm as a case study. Its computational methodology is comparatively straightforward. Furthermore, it exhibits minimal demands on hardware resources, thereby facilitating its implementation on devices characterized by constrained computational capabilities, such as basic communication terminal devices.
2. **Effective Suppression of Stationary Interference:** In scenarios where the statistical properties of the interference signal (for instance, PSD) exhibit relative stability over a defined temporal duration, time-domain adaptive filtering methods are capable of progressively adjusting to the interference signal by modifying the filter coefficients, thereby achieving a significant reduction in the interference. For instance, in a communication environment characterized by a relatively invariant source of interference, such methodologies can gradually acquire knowledge of the interference characteristics and mitigate its detrimental effects.

   A profound comprehension of the convergence attributes is essential. The rate of convergence is primarily governed by the step-size parameter (μ), and the impact of this parameter on convergence is comparatively straightforward. Through the modulation of the step-size parameter, the velocity of convergence for the filter weights can be regulated to a certain degree. Academics and practitioners in the field can adeptly modify the convergence mechanism in accordance with particular interference scenarios and system specifications.

Disadvantages

1. **Inadequate Adaptability to Nonstationary Interference:** In scenarios where the statistical properties of the interference signal exhibit rapid temporal fluctuations, time-domain adaptive filtering may encounter difficulties with slow tracking. This phenomenon arises because the weight adjustment is contingent upon the error at the present symbol instant; thus, in instances of rapid interference variation, the filter may struggle to promptly adjust to the new interference characteristics. For instance, within a complex and dynamic electromagnetic environment characterized by multiple interference sources with continually varying frequencies and amplitudes, time-domain adaptive filtering may prove to be ineffective.
2. **The Trade-off of the Step-size Parameter:** The selection of the step-size parameter μ embodies a significant trade-off. On one hand, an elevated value of μ facilitates a more rapid convergence of the weights, thereby allowing for the swift tracking and mitigation of fluctuating interference. Conversely, an increased value of μ results in an augmented variance of the weights, which adversely impacts the efficacy of interference mitigation. This trade-off

presents challenges in identifying an optimal step-size parameter in practical applications that fulfill the dual requirements of convergence velocity and interference suppression efficacy.

6.4.3.2 Transform-domain Adaptive Filtering

Advantages

1. **Rapid Tracking of Interference Fluctuations:** It possesses the capability to swiftly track the fluctuations of narrow-band jamming without necessitating an iterative approach. This characteristic renders it particularly adept at addressing swiftly evolving interference signals. For instance, in a dynamic communication environment where the frequency and amplitude of the interference signal undergo frequent alterations, transform-domain adaptive filtering can rapidly adapt to these modifications and efficiently mitigate the adverse effects of interference on the signal.
2. **Efficient Management of Narrow-band Interference:** By conducting an analysis of the received signal within the frequency domain, it is feasible to directly address narrow-band interference. Through methodologies such as interference detection and band-pass filtering, it is capable of effectively mitigating narrowband interference. For instance, a communication system characterized by multiple sources of narrow-band interference can precisely identify and filter these disruptive signals.

Disadvantages

1. **Elevated Computational Complexity:** The necessity of converting the signal between the time domain and the frequency domain, coupled with intricate processing within the frequency domain (including tasks such as interference identification and band-pass filtering), typically results in augmented computational complexity in comparison to time-domain adaptive filtering methodologies. Consequently, this may necessitate the utilization of more robust computing resources to facilitate the operational demands of the algorithm, thereby rendering it less feasible for implementation on devices with limited resources.
2. **Limited Ability to Manage Broadband Interference:** While it possesses merits in addressing narrow-band interference, its efficacy in processing broadband interference may not reach the same level as that for narrow-band interference. This is attributed to the more intricate spectral characteristics associated with broadband interference, rendering identification and processing in the frequency domain considerably more challenging. For instance, in the presence of broadband noise interference, it may not achieve the same degree of precision in filtering out the interference components as it does when confronting narrow-band interference.

6.5 AI-based Solutions

Traditional methodologies for mitigating interference, such as DSSS and FHSS, exhibit a degree of resilience against certain types of interference. Nonetheless, as communication technologies advance and the complexity of the interference landscape increases, the inherent limitations of these techniques become evident. In the face of novel interference types, these methodologies struggle to adequately fulfill the requirements for effective communication countermeasures. Within this framework, AI technology, characterized by its robust learning capabilities, analytical prowess, and adaptability, offers a promising avenue for enhancing communication interference mitigation. It possesses the ability to extract valuable information from extensive signal datasets, swiftly and accurately address interference challenges, and has emerged as a pivotal technology in resolving the intricate issues associated with communication anti-interference.

6.5.1 Method Descriptions

6.5.1.1 Interference Signal Identification and Classification

Deep learning (DL) algorithms, including convolutional neural networks (CNNs) and recurrent neural networks (RNNs), have been employed to assess the acquired signals. CNNs demonstrate proficiency in the processing of data characterized by spatial structures and are capable of efficiently extracting signal features to discern various categories of interference signals, such as narrow-band and wide-band interference. RNNs exhibit strong performance in the analysis of time-series data, enabling them to scrutinize the evolving patterns of interference signals over time, thereby facilitating a more precise classification of such signals. Through the training on a substantial dataset of interference signal samples, the model is positioned to accurately identify interference signals, thereby establishing a foundation for the development of subsequent anti-interference strategies.

6.5.1.2 Adaptive Anti-Interference Strategy Generation

By integrating reinforcement learning (RL) methodologies, the communication system is enabled to autonomously acquire and develop the most effective anti-interference strategies in accordance with the nature and magnitude of interference signals. The communication system functions as an intelligent agent. During its interactions with the interference environment, it continually engages in diverse anti-interference operations (such as modulation of communication frequency, adjustment of power levels, and variation of coding techniques, among others) and, based on the reward signals provided by the environment (such as enhancements in communication quality and reductions in BER, etc.), it acquires the optimal strategy. For instance, upon the identification of a substantial interference signal, the RL model is capable of promptly modifying the communication frequency and transitioning to a frequency band characterized by diminished interference, thereby safeguarding the continuity of communication.

6.5.1.3 Interference Prediction and Advance Prevention

ML algorithms are employed to scrutinize historical interference data and formulate predictive models for interference. For instance, the autoregressive integrated moving average (ARIMA) model, grounded in time series analysis, is capable of forecasting the probability of occurrence, intensity, and frequency of interference signals within a specified future timeframe based on prior interference occurrences. The communication system preemptively modifies its communication parameters or implements corresponding anti-interference strategies in accordance with the predictive outcomes, thereby enabling proactive measures to mitigate interference and enhancing the resilience of the communication system against such disruptions.

6.5.1.4 Multi-agent Cooperative Anti-interference

In a multifaceted communication milieu characterized by numerous communication nodes, the implementation of multi-agent RL is proposed. Each communication node functions as an autonomous intelligent agent, engaging in collaborative efforts to mitigate interference. Each intelligent agent possesses the capability to disseminate interference-related data and strategies for counteracting interference, thereby optimizing the allocation of communication resources through collective action to enhance the overall performance against interference. For instance, when a specific node identifies the presence of interference, it can relay the interference data to other nodes, which in turn can modify their communication parameters based on their individual circumstances to avert being adversely affected by the same interference.

6.5.2 Applications

In light of prevailing uncertainties, AI-assisted active security enhancement methodologies can be effectively employed for both predictive analytics and detection purposes. Additionally, these methodologies may be utilized as a highly effective passive security approach to address intricate optimization challenges. A prominent domain of application for AI pertains to the resolution of beam-hopping dilemmas associated with LEO SCSs, with the primary aim of mitigating detrimental interference effects. Xu et al. conducted an investigation into deep RL-aided dynamic beam hopping within multi-beam satellite frameworks [17].

Specifically, to address the challenges posed by erratically varying traffic demands and time-dependent wireless channel conditions, deep RL was integrated with simulated annealing techniques as reported in [18]. Furthermore, multi-objective deep RL methodologies were introduced in [19]. Additionally, in [20], a multi-agent deep RL framework was proposed, wherein each beam functioned as an individual agent while operating in a cooperative manner. The objectives and constraints associated with the formulated beam-hopping optimization problem are typically linked to the SINR, which is directly affected by the CCI encountered. To alleviate CCI among neighboring beams, the intelligent approach developed in [20] estimated the signal strengths encountered in overlapping regions and systematically organized the spatial arrangement of the beams to ensure that adjacent beams do not utilize identical frequency resources.

In [21], the challenges associated with interference-related sum-rate maximization have been articulated with the intent of enhancing the optimization of the passive BF vector at the RIS, which were resolved utilizing graph attention networks that fall within the classification of unsupervised offline learning methodologies. These networks demonstrated proficiency in accurately capturing the dynamic topology of RIS-assisted LEO SCSs while maintaining a low online computational complexity. Furthermore, the authors of [22] employed a DL framework to ascertain the optimal RIS coefficients within the context of maximizing secrecy capacity, thereby achieving the desired long-term objectives with high convergence efficiency and sample efficiency.

On the contrary, sophisticated passive security enhancement mechanisms may be employed to thwart jamming assaults targeting LEO SCSs. As wireless networks evolve in complexity, so too do the corresponding jamming strategies. For instance, certain jamming actions may exhibit capabilities akin to learning and reasoning processes. The SCS that operates under conditions of cyclic visibility and fixed orbital parameters is particularly vulnerable to these advanced jamming techniques, which could result in significant congestion. Conventional anti-jamming strategies, including DSSS, FHSS, multibeam antennas, and self-adaptive routing protocols, are inadequate in effectively countering these sophisticated jamming methods. A collection of innovative anti-jamming frameworks has been devised for SCSs by Han et al. Specifically, in [23], they articulated a hierarchical anti-jamming Stackelberg game to elucidate the dynamics between intelligent jammers and satellite users. Additionally, they introduced a two-tiered anti-jamming protocol, wherein the initial tier employs deep RL to diminish the routing decision space. In contrast, the subsequent tier utilizes Q-learning to expedite anti-jamming routing processes. In their subsequent research [24], the authors established a distributed dynamic anti-jamming mechanism tailored for a satellite-supported military IoT network, achieving a reduction in energy consumption without significantly compromising performance metrics. The nature of the jamming attacks was thoroughly examined, and countermeasures were implemented through deep RL-informed anti-jamming strategies. Nevertheless, as their methodology was predicated on the assessment of the adversarial interactions between jammers and legitimate users, the efficacy of the anti-jamming measures was invariably contingent upon the precision of jamming attack detection. More recently, the authors of [25] introduced a cross-layer anti-jamming strategy that encompasses both the link layer and the network layer, demonstrating superior performance compared to monocular anti-jamming approaches. The link layer addressed specific channel jamming through the application of advanced Q-learning techniques, while the network layer mitigated ISL jamming by identifying alternative routing paths utilizing a deep Q network algorithm. However, this algorithm is predominantly applicable to jamming issues within low-dimensional contexts, constrained by limited processing capabilities and energy expenditures, thereby leaving a plethora of unresolved challenges for future exploration.

AI has achieved significant advancements in the protection of LEO SCSs from both active and passive information domain perspectives. The distinctive characteristics and evolutionary patterns of traffic can be precisely forecasted by AI methodologies, thereby facilitating the mitigation of potential congestion and elevated CCI. Furthermore, AI can function as a near-optimal solution provider for intricate non-convex optimization challenges encountered by hybrid SCSs within beam-hopping frameworks, RISs, and anti-jamming methodologies, among others. Nevertheless, the efficacy of AI implementations in LEO SCSs is contingent upon the availability of a sufficiently extensive dataset for training, which may be unattainable in real-world communication contexts.

6.5.3 Advantages and Disadvantages

In contemporary communication frameworks, interference has emerged as a pervasive obstacle that profoundly compromises the integrity and dependability of communication. Whether it pertains to the uninterrupted streaming of high-definition videos on contemporary 5G-enabled mobile devices, the consistent functionality of industrial IoT within intricate manufacturing settings, or vital communication necessary for aerospace and military operations, interference has the propensity to disrupt data transmission, precipitate communication failures, and engender considerable economic detriments alongside potential safety hazards.

Traditional methodologies for counteracting interference exhibit inherent constraints when addressing the progressively intricate and varied sources of disruption. Consequently, AI has surfaced as a viable and innovative approach within the domain of interference mitigation, capitalizing on its distinctive capabilities to transform the strategies employed against interference. At this juncture, it is imperative to conduct a thorough examination of the merits and demerits associated with the implementation of AI in the context of interference management.

Advantages

1. **Strong Adaptability:** In 5G communication, AI has the capacity to dynamically modify transmission parameters in response to surrounding interference, thereby facilitating the seamless playback of high-definition video content. When confronted with complex and variable interference, such as that produced by diverse devices within an industrial environment, AI can rapidly adjust and enhance communication strategies.
2. **Excellent Signal Processing:** CNN and RNN possess the capability to accurately discern interference signals. For instance, within the context of satellite communication, these networks can extract critical features from intricate signals, including cosmic ray interference and mining interference patterns, thereby offering insights for subsequent anti-interference measures.
3. **Good Multiagent Cooperation:** Within the vehicle-to-everything (V2X) network, each vehicle node functions as an autonomous intelligent agent. These agents collaborate with one another, disseminate interference information, optimize resource distribution, effectively mitigate signal congestion, and enhance the overall efficacy of anti-interference mechanisms.
4. **Predictive Ability:** In smart grid communication, AI scrutinizes historical interference data, forecasts potential interference during thunderstorm conditions, and implements preemptive measures to ensure the stability of power communication systems.

Disadvantages

1. **High Data Requirements:** The procurement of a substantial volume of high-quality labeled data presents significant challenges. For instance, the acquisition of data pertaining to novel categories of malicious interference signals is particularly arduous. The presence of subpar data may result in erroneous conclusions drawn by the model in intricate scenarios, thereby undermining the efficacy of anti-interference measures.
2. **High Resource Cost:** DL algorithms impose considerable demands on computational resources, which may be prohibitive for small-scale IoT devices.

The processes of model training and optimization are protracted and time-intensive. In the context of emergency communication, timely adjustments to strategies in response to abrupt interference are rendered impractical.

3. **Poor Model Interpretability:** The decision-making mechanisms of neural networks are often characterized as a "black box." In financial communication, elucidating the rationale behind decision-making is exceedingly challenging. When issues arise, the processes of troubleshooting and remediation encounter substantial impediments.

References

1 Xiaojin Ding, Tiecheng Song, Yulong Zou, et al. Security-reliability tradeoff analysis of artificial noise aided two-way opportunistic relay selection. *IEEE Trans. Veh. Technol.*, 66(5):3930–3941, May 2017.

2 Yansha Deng, Lifeng Wang, Syed Ali Raza Zaidi, et al. Artificial-noise aided secure transmission in large scale spectrum sharing networks. *IEEE Trans. Commun.*, 64(5):2116–2129, May 2016.

3 Sangseok Yun, Jae-Mo Kang, Il-Min Kim, et al. Deep artificial noise: Deep learning-based precoding optimization for artificial noise scheme. *IEEE Trans. Veh. Technol.*, 69(3):3465–3469, Mar. 2020.

4 Yi Liao, Jian Wang, and Qing Huo Liu. Transmit beampattern synthesis for frequency diverse array with particle swarm frequency offset optimization. *IEEE Trans. Antennas Propag.*, 69(2):892–901, Feb. 2021.

5 Wenqin Wang. Frequency diverse array antenna: New opportunities. *IEEE Antennas Propag. Mag.*, 57(2):145–152, Apr. 2015.

6 Yi Liao, Wenqin Wang, and Huaizong Shao. Symmetrical logarithmic frequency diverse array for target imaging. In *Proceedings of the IEEE Radar Conference*, pages 0039–0042, Oklahoma City, OK, USA, 23–27 Apr. 2018.

7 Kuandong Gao, Jingye Cai, and Jie Xiong. Decoupled frequency diverse array range–angle-dependent beampattern synthesis using non-linearly increasing frequency offsets. *IET Microw. Antennas Propag.*, 10:880–884, Jun. 2016.

8 Jinsong Hu, Shihao Yan, Feng Shu, Jiangzhou Wang, Jun Li, and Yijin Zhang. Artificial-noise-aided secure transmission with directional modulation based on random frequency diverse arrays. *IEEE Access*, 5:1658–1667, Jan. 2017.

9 Paolo Rocca, Randy L Haupt, and Andrea Massa. Interference suppression in uniform linear arrays through a dynamic thinning strategy. *IEEE Trans. Antennas Propag.*, 59(12):4525–4533, Dec. 2011.

10 Pingyue Yue, Jianping An, Jiankang Zhang, et al. Low earth orbit satellite security and reliability: Issues, solutions, and the road ahead. *IEEE Commun. Surv. Tutor.*, 25(3):1604–1652, 2023.

11 Tapan Kumar Sarkar, Hong Wang, Sheeyun Park, et al. A deterministic least-squares approach to space-time adaptive processing (STAP). *IEEE Trans. Antennas Propag.*, 49(1):91–103, Jan. 2001.

12 Diego Cristallini and Wolfram Burger. A robust direct data domain approach for STAP. *IEEE Trans. Signal Process.*, 60(3):1283–1294, Mar. 2012.

13 Scott C Douglas, Quanhong Zhu, and Kent F Smith. A pipelined LMS adaptive FIR filter architecture without adaptation delay. *IEEE Trans. Signal Process.*, 46(3):775–779, Mar. 1998.

14 Khaled Mayyas. Performance analysis of the deficient length LMS adaptive algorithm. *IEEE Trans. Signal Process.*, 53(8):2727–2734, Aug. 2005.

15 Balaji Raghothaman, Darel Linebarger, and Dinko Begusic. A new method for low rank transform domain adaptive filtering. *IEEE Trans. Signal Process.*, 48(4):1097–1109, May 2000.

16 Zhangsheng Wang, Ming Lv, and Bin Tang. Paper application of partial coefficient update LMS algorithm to suppress narrowband interference in DSSS system. In *Proceedings of the International Conference on Communication Software and Networks*, pages 275–278, Chengdu, China, 27–28 Feb. 2009.

17 Yuchen Zhang, Xin Hu, Rong Chen, et al. Dynamic beam hopping for DVB-S2X satellite: A multi-objective deep reinforcement learning approach. In *2019 IEEE International Conferences on Ubiquitous Computing & Communications (IUCC) and Data Science and Computational Intelligence (DSCI) and Smart Computing, Networking and Services (SmartCNS)*, pages 164–169, Shenyang, China, 21–23 Oct. 2019. IEEE.

18 Xin Hu, Yin Wang, Zhijun Liu, et al. Dynamic power allocation in high throughput satellite communications: A two-stage advanced heuristic learning approach. *IEEE Trans. Veh. Technol.*, 72(3):3502–3516, Apr. 2023.

19 Xin Hu, Yuchen Zhang, Xianglai Liao, et al. Dynamic beam hopping method based on multi-objective deep reinforcement learning for next generation satellite broadband systems. *IEEE Trans. Broadcast.*, 66(3):630–646, Sep. 2020.

20 Xin Hu, Xianglai Liao, Zhijun Liu, et al. Multi-agent deep reinforcement learning-based flexible satellite payload for mobile terminals. *IEEE Trans. Veh. Technol.*, 69(9):9849–9865, Sep. 2020.

21 Hailin Cao, Wang Zhu, Wenjuan Feng, et al. Robust beamforming based on graph attention networks for IRS-assisted satellite IoT communications. *Entropy*, 24(3):326, Feb. 2022.

22 Chong Huang, Gaojie Chen, Yitong Zhou, et al. Deep learning empowered secure RIS-assisted non-terrestrial relay networks. In *Proceedings of the Vehicular Technology Conference*, pages 1–5, London, UK, Sep. 2022.

23 Chen Han, Liangyu Huo, Xinhai Tong, et al. Spatial anti-jamming scheme for internet of satellites based on the deep reinforcement learning and Stackelberg game. *IEEE Trans. Veh. Technol.*, 69(5):5331–5342, May 2020.

24 Chen Han, Aijun Liu, Haichao Wang, et al. Dynamic anti-jamming coalition for satellite-enabled army IoT: A distributed game approach. *IEEE Internet Things J.*, 7(11):10932–10944, Nov. 2020.

25 Peijie Yan, Feihuang Chu, Luliang Jia, et al. A cross-layer anti-jamming method in satellite internet. *IET Commun.*, 17(1):121–133, Oct. 2022.

7

Security-related Reliability Enhancement Solutions

7.1 Secrecy Capacity/Throughput

7.1.1 Definitions

The secrecy capacity sets the fundamental limit of secure communications over noisy channels. It functions as the counterpart to the usual point-to-point channel capacity but with an additional information-theoretic secrecy requirement. In contrast to traditional communication channels, where the paramount concern is reliability, secrecy capacity considers both reliable communication and confidentiality, which entails keeping information secret from eavesdroppers while maintaining successful transmission between legitimate parties.

Secrecy capacity is inherently associated with the wiretap channel model. The concept of wiretap channels was introduced in [1] and has since been a fundamental model in information-theoretic security. Specifically, a wiretap channel consists of three parties:

- Transmitter (typically referred to as Alice): The party that intends to send a secure message.
- Legitimate receiver (typically referred to as Bob): The intended recipient of the secure message.
- Eavesdropper (typically referred to as Eve): The unauthorized party attempting to intercept the message.

In this model, Alice sends a message over a legitimate channel to the legitimate receiver Bob, while the eavesdropper Eve tries to intercept the message over a different, possibly degraded, eavesdropping channel. The primary objective of PLS is to guarantee that Bob is capable of reliably decoding the message, while concurrently preventing Eve from acquiring any pertinent information regarding the transmitted content. Different from the performance evaluation metrics in typical

Safeguarding Satellite Communications: Issues, Challenges, and Solutions, First Edition.
Jianping An, et al.

communication systems, the target of PLS is twofold, i.e., the reliability of Bob's decoding and the confidentiality against Eve's eavesdropping.

According to Wyner's theory of eavesdropping channels, the secrecy capacity of the Alice–Bob–Eve model can be expressed as

$$C_S = [\log(1+\gamma_B) - \log(1+\gamma_E)]^+, \tag{7.1}$$

where γ_B and γ_E represent the SNR at Bob and Eve, respectively, and $[\cdot]^+ \triangleq \max\{\cdot, 0\}$; $\log(1+\gamma_B)$ and $\log(1+\gamma_E)$ denote the channel capacity of Alice–Bob and Alice–Eve links, respectively.

The secrecy capacity C_S represents the maximum rate at which information can be sent from Alice to Bob without being decoded by Eve. It captures the idea that secrecy can only be maintained if Bob can receive the information more effectively than Eve. Apparently, there are two possible cases:

- If $\log(1+\gamma_B) > \log(1+\gamma_E)$: This means Bob's channel capacity is greater than Eve's. The difference $\log(1+\gamma_B) - \log(1+\gamma_E)$ is positive, meaning there exists a positive rate at which information can be sent securely.
- If $\log(1+\gamma_B) \leq \log(1+\gamma_E)$: This means Eve's channel capacity is equal to or greater than Bob's. The difference $\log(1+\gamma_B) - \log(1+\gamma_E)$ therefore is zero or negative. To prevent a negative secrecy capacity, which does not make sense, the $[\cdot]^+$ operator is used, which sets any negative value to zero. In this case, the security of the transmission breaks down, making it impossible to ensure that Bob gets the message securely without Eve potentially intercepting it. Specifically, there are two possible outcomes: Firstly, Eve can eavesdrop on all the information Alice transmits to Bob, gaining access to the entire message without any extra effort; secondly, Bob struggles to reliably decode the information from Alice.

Furthermore, secrecy throughput serves as a more holistic metric for evaluating the efficiency of secure communication. It assesses the volume of secure data that can be transmitted over a specified duration, integrating the theoretical security measure (secrecy capacity) with practical efficiency considerations. Typically, secrecy throughput can be presented as

$$T_S = C_S \cdot \eta, \tag{7.2}$$

where $\eta \in (0, 1)$ is the utilization factor that describes the efficiency of the communication channel, considering the real-world inefficiencies such as noise, protocol overhead, error rates, and other system limitations.

In conclusion, secrecy capacity denotes the theoretical maximum rate at which information can be transmitted securely from a sender to a receiver, devoid of interception by an eavesdropper, presuming ideal system conditions while disregarding practical constraints. Conversely, secrecy throughput advances this concept by accounting for the real-world inefficiencies that characterize the degree of channel utilization, which is impacted by factors such as protocol overhead and error rates, thereby offering a more pragmatic assessment of the actual amount of secure data that is transmitted successfully. Consequently, while secrecy capacity embodies the potential for secure communication, secrecy throughput delineates the effective and attainable rate of secure communication in practical scenarios, achieving a balance between security and efficiency.

7.1.2 Secrecy Capacity/Throughput in Satellite Communications

In satellite communication networks, security is profoundly critical due to their intrinsic broadcast characteristics, which render them vulnerable to eavesdropping and unauthorized access. The physical layer, encompassing the transmission and reception of signals, is especially susceptible to both passive and active attacks. For instance, signals transmitted from satellites typically cover large terrestrial areas and can be intercepted by illegitimate parties with powerful equipment in Figure 7.1. This possibility of interception underscores the necessity for robust PLS measures, such as encryption and signal obfuscation, to safeguard sensitive information and ensure that only designated recipients possess the capability to decode the transmitted data. Furthermore, PLS can be integrated with alternative security methodologies, including SS and FH, to mitigate various categories of attacks, such as jamming and interference, which have the potential to disrupt communication services and compromise the integrity of the conveyed data. Consequently, the exploration of advanced security protocols at the physical layer is imperative for the protection of satellite communication networks against a diverse spectrum of potential threats.

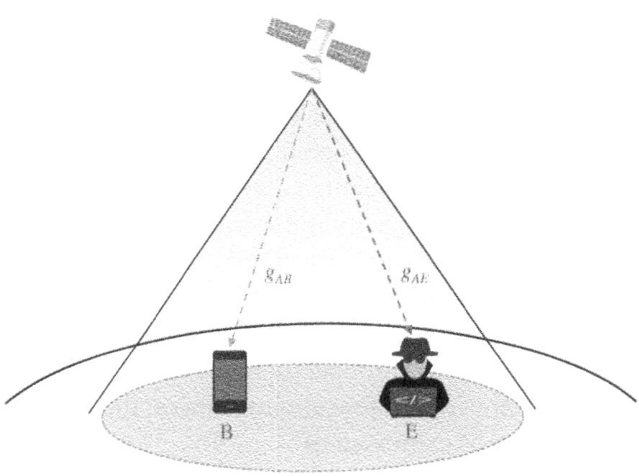

Figure 7.1 Eavesdropping scenario in space-terrestrial downlink communications.

To comprehensively evaluate the secrecy performance of satellite networks, it is imperative to first examine the characteristics of SNR, as SNR constitutes a pivotal metric that affects various performance parameters within SCSs [2]. SNR measures the quality of the transmitted signal in relation to ambient noise, which directly influences the overall performance and reliability of the system. An elevated SNR signifies a more distinct and robust signal relative to noise, thereby facilitating enhanced performance across multiple dimensions of the communication framework.

In order to evaluate the characteristics of SNR in satellite communications, we initially delineate the attributes of space-terrestrial wireless communication channels. Numerous methodologies exist for the modeling of space-terrestrial channels. A precise analytical framework was developed by [3], which is recognized as the shadowed-Rician fading model. Specifically, under this channel model, the small-scale fading factor can be written as

$$\begin{aligned} f_g(x) = {} & \left(\frac{2b_S m_S}{2b_S m_S + \Omega_S}\right)^{m_S} \frac{1}{2b_S} \exp\left(-\frac{x}{2b_S}\right) \\ & \times {}_1F_1\left(m_S; 1; \frac{\Omega_S x}{2b_S\left(2b_S m_S + \Omega_S\right)}\right), \end{aligned} \tag{7.3}$$

where $g \triangleq |h|^2$, h is the channel coefficient characterizing the space-terrestrial channel, $2b_S$, m_S, and Ω_S denote the average power of the scatter component, fading severity of the line-of-sight component, and the average power of the line-of-sight component, respectively, ${}_1F_1(\cdot;\cdot;\cdot)$ represents the confluent hypergeometric function of the first kind.

Based on the channel properties, the SNR of space-terrestrial communications can be expressed as

$$\gamma = \frac{Pgd^{-\alpha}}{\sigma^2}, \tag{7.4}$$

where P is the transmit power, σ^2 is the additive white Gaussian noise (AWGN) at the receiver, d is the propagation distance of the signal, and α is the path loss exponent determined by the propagation environment.

Thus, returning to the definitions of secrecy capacity and secrecy throughput, we are able to formulate these metrics in satellite communication networks accordingly. Specifically, we consider a space-terrestrial downlink communication scenario, in which a satellite (denoted as A) transmits to a legitimate ground destination (denoted as B), while an illegitimate eavesdropper (denoted as E) endeavors to intercept the transmitted information. Under these circumstances, the secrecy capacity of this communication link can be expressed as

$$\begin{aligned} C_S &= \left[\log(1+\gamma_B) - \log(1+\gamma_E)\right]^+ \\ &= \left[\log\left(1 + \frac{P_A g_{AB} d_{AB}^{-\alpha_{AB}}}{\sigma_B^2}\right) - \log\left(1 + \frac{P_A g_{AE} d_{AE}^{-\alpha_{AE}}}{\sigma_E^2}\right)\right]^+, \end{aligned} \tag{7.5}$$

where P_A is the transmit power at satellite A; $g_{AB} = |h_{AB}|^2$ and $g_{AE} = |h_{AE}|^2$ represent the channel power gain of A–B and A–E links, respectively; σ_B^2 and σ_E^2 represent the AWGN average power at B and E.

Besides, the secrecy throughput of this downlink scenario can be written as

$$T_S = C_S\eta = \eta\left[\log\left(1 + \frac{P_A g_{AB} d_{AB}^{-\alpha_{AB}}}{\sigma_B^2}\right) - \log\left(1 + \frac{P_A g_{AE} d_{AE}^{-\alpha_{AE}}}{\sigma_E^2}\right)\right]^+. \quad (7.6)$$

7.1.3 Illustrative Results

To reflect the expected secrecy capacity under small-scale fading on space-terrestrial links, average secrecy capacity (ASC) is further adopted to evaluate the secrecy performance. Specifically, the ASC can be expressed as follows:

$$C_S^{(A)} = \int_0^\infty \int_0^\infty C_S(g_{AB}, g_{AE}) f_{g_{AB}}(g_{AB}) f_{g_{AE}}(g_{AE})\, dg_{AB} dg_{AE}. \quad (7.7)$$

The secrecy capacity C_S is a function of channel conditions g_{AB} and g_{AE}, determined by the relative strength of the legitimate and eavesdropping channels. By substituting (7.5) and (7.3) into (7.7), the ASC can be calculated through averaging over g_{AB} and g_{AE}.

The ASC is illustrated in Figure 7.2. It is evident that the ASC increases through the amplification of the satellite's transmit power. Furthermore, a reduced propagation distance d_{AB} enhances the SNR at the legitimate receiver, thereby augmenting the performance of the secrecy capacity.

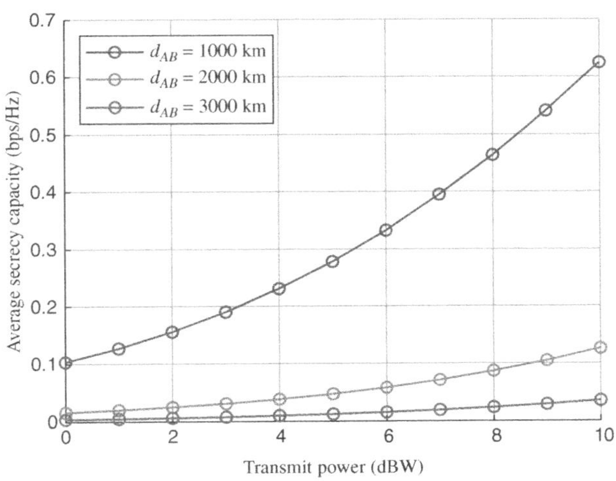

Figure 7.2 ASC with respect to transmit power and propagation distance.

7.2 Outage/BER/Packet Error Rate (PER)

7.2.1 Definitions

SOP serves as a quantitative metric within the context of PLS to measure the probability that the secrecy capacity of a communication channel may descend below a specified threshold, thereby leading to the inability to uphold secure communication [4]. In essence, it measures the probability that Eve can successfully intercept and decode the information being transmitted, which ultimately undermines the security of the communication established between the legitimate users, Alice and Bob.

To be specific, the SOP can be written as follows:

$$P_{\mathrm{out}} = \Pr\{C_S < R_S\}, \tag{7.8}$$

wherein R_S represents the rate of the secrecy information adopted by Alice, which serves as the secrecy capacity threshold.

A diminished SOP signifies an elevated degree of security, as it implies a reduced probability of an eavesdropper effectively seizing the communication. Comprehending and mitigating SOP is imperative for the formulation of secure communication frameworks, particularly in satellite communication contexts characterized by extensive coverage areas and intricate eavesdropping conditions.

Secrecy capacity and secrecy throughput are theoretical concepts that define the potential for secure communication. However, ideal conditions and asymptotic behavior are assumed for these two metrics. In contrast, BER and PER are practical metrics that evaluate the actual performance of secure communication systems. Thus, it is essential to take BER and PER performance into account when evaluating the performance of such communication networks.

By definition, BER measures the number of erroneous bits received compared to the total number of bits sent, and PER measures the number of erroneous packets received compared to the total number of packets sent. BER is a direct indicator of the quality of the received signal, while PER is particularly useful for protocols that transmit data in packets, such as slotted ALOHA [5]. The importance of these two metrics in PLS can be summarized as follows:

- BER and PER are crucial measures of the quality of received signals, indicating the current channel state. As such, they can be utilized in dynamic transmission schemes that implement adaptive modulation, coding, and power control to maintain the desired secrecy capacity/throughput.
- In communication architectures that utilize packetized frameworks, elevated PER may result in recurrent retransmission occurrences, thereby inducing latency and augmenting the likelihood for an eavesdropper to analyze and potentially crack the transmitted data. Consequently, BER and PER are intricately linked to the confidentiality of the communication system.

- BER and PER also serve as critical indicators for the effectiveness of error correction and detection schemes employed in PLS. By monitoring these metrics, the system can dynamically adjust error control mechanisms to enhance the reliability and security of the communication [6]. High BER or PER values can trigger stronger ECCs, while lower values can allow for more efficient transmission, thereby achieving a balance between security imperatives and performance optimization.

Moreover, finite blocklength (FBL) communications with prospective applications in URLLC scenarios employ a blocklength error rate (BLER) for performance characterization. According to [7], the capacity in the FBL regime can be expressed as

$$C = \log_2 (1+\gamma) - \sqrt{\frac{V(\gamma)}{N}} Q^{-1}(\epsilon), \tag{7.9}$$

where $\gamma = \frac{P}{N_0}$ is the SNR at the receiver side, $V(\gamma) = \ln^2 2 \left(1 - (1+\gamma)^{-2}\right)$ denotes the channel dispersion, N is the blocklength, ϵ is the BLER of the transmitted block, and $Q^{-1}(\cdot)$ represents the inverse Q-function. Based on this formula, for a given target rate R, the BLER metric can be obtained as

$$\epsilon = Q\left(\sqrt{\frac{N}{V(\gamma)}} \left(\log_2 (1+\gamma) - R\right)\right). \tag{7.10}$$

By substituting (7.4) into (7.10), the BLER in a satellite communications setting can be derived.

7.2.2 Outage/BER/PER in Satellite Communications

Without loss of generality, we consider the aforementioned scenario of space-terrestrial downlink transmission. Similar to the previous analysis of SNR, the SOP can be specifically expressed as

$$\begin{aligned}
P_{\text{out}} &= \Pr\{C_S < R_S\} \\
&= \Pr\left\{\left[\log\left(1 + \frac{P_A g_{AB} d_{AB}^{-\alpha_{AB}}}{\sigma_B^2}\right) - \log\left(1 + \frac{P_A g_{AE} d_{AE}^{-\alpha_{AE}}}{\sigma_E^2}\right)\right]^+ < R_S\right\} \\
&= \Pr\left\{g_{AB} < \frac{\sigma_B^2 d_{AB}^{\alpha_{AB}}}{P_A}\left[2^{R_S}\left(1 + \frac{P_A g_{AE} d_{AE}^{-\alpha_{AE}}}{\sigma_E^2}\right) - 1\right]\right\} \\
&= \int_0^\infty F_{g_{AB}}\left(\frac{\sigma_B^2 d_{AB}^{\alpha_{AB}}}{P_A}\left[2^{R_S}\left(1 + \frac{P_A y d_{AE}^{-\alpha_{AE}}}{\sigma_E^2}\right) - 1\right]\right) f_{g_{AE}}(y) dy,
\end{aligned} \tag{7.11}$$

where $F_{g_{AB}}(\cdot)$ and $f_{g_{AE}}(\cdot)$ denote the cumulative distribution function (CDF) and probability density function (PDF) of g_{AB} and g_{AE}, respectively.

On the other hand, BER and PER at the legitimate receiver depend on the specific modulation and/or medium access control schemes. Taking binary phase shift keying (BPSK) modulation as an example, the BER at B can be formulated as

$$\mathrm{BER}_B = Q\left(\sqrt{\frac{2P_A g_{AB}}{\sigma_B^2}}\right). \tag{7.12}$$

Assuming independent bit errors inside each packet, the PER can be expressed as follows, as any bit error will result in information loss of the whole packet:

$$\mathrm{PER}_B = 1 - (1 - \mathrm{BER}_B)^N, \tag{7.13}$$

where N represents the bit length of each packet.

It is noted that there are different approaches to improving the reliability of legitimate transmission, which can be observed from the expression of the received SNR at B. Specifically, $\frac{P_A g_{AB} d_{AB}^{-\alpha_{AB}}}{\sigma_B^2}$ can be increased by increasing P_A or g_{AB}, which means lower SOP for a specific channel condition of E and lower BER at B according to (7.12). Hence, raising transmit power P_A or improving channel condition g_{AB} by BF, equalization, etc., can scale up the security and reliability level simultaneously.

7.2.3 Illustrative Results

The influence of transmission power and propagation distance on the SOP is illustrated in Figure 7.3. It is ascertained that the SOP diminishes as the transmission power at the satellite increases. Furthermore, the propagation distance exerts a

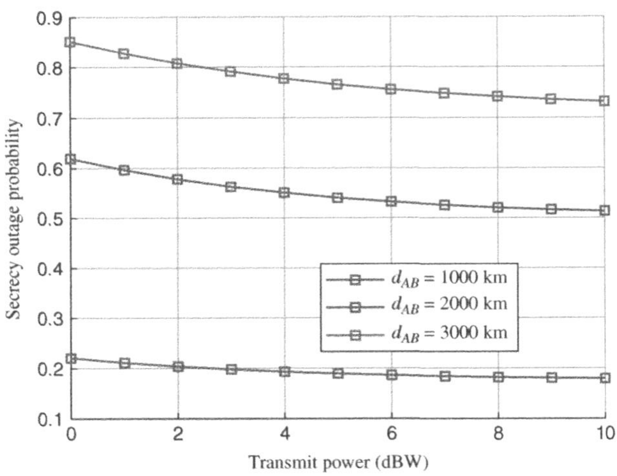

Figure 7.3 SOP with respect to transmit power and propagation distance.

similar impact on the SOP as the ASC, thereby necessitating meticulous design considerations for satellite orbits to augment the security level.

Furthermore, the impact of transmission power and propagation distance on the BER and PER at the legitimate user is illustrated in Figures 7.4 and 7.5, respectively. It is discernible that an augmentation in transmit power enhances the

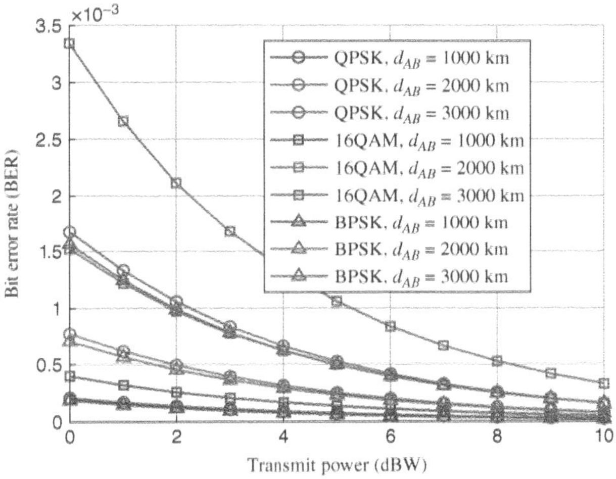

Figure 7.4 BER with respect to transmit power and propagation distance.

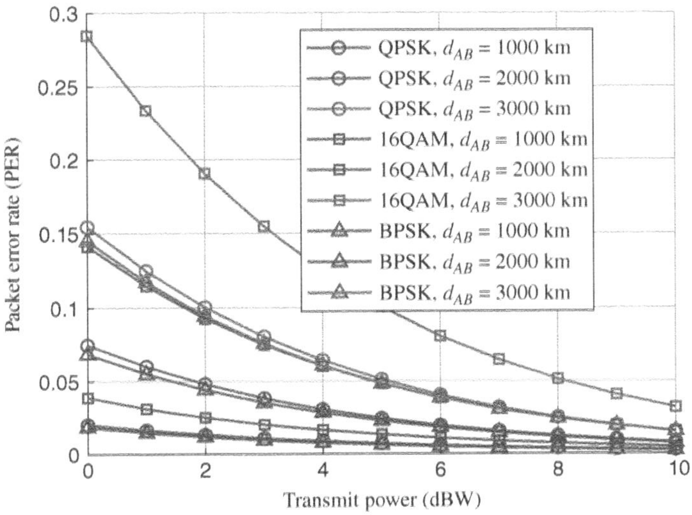

Figure 7.5 PER with respect to transmit power and propagation distance.

performance of BER and PER, thereby increasing the reliability of the legitimate communication link. Moreover, a reduced propagation distance of the legitimate link is advantageous for its reliability.

7.3 AI-based Solutions to Collisions and Debris

7.3.1 Collision Avoidance

The accelerated proliferation of mega-constellations within LEO has markedly heightened the probability of collision incidents. LEO is densely populated with a diverse array of objects, encompassing components of rocket bodies, inactive satellites, fragments, and debris, all of which traverse at elevated velocities in an uncontrolled manner. Collisions occurring in this milieu can result in the generation of additional debris, thereby potentially instigating a cascade of subsequent collisions—a phenomenon commonly known as the Kessler Syndrome. This escalating risk accentuates the pressing necessity for augmented collision avoidance strategies to be implemented by space organizations and agencies.

7.3.1.1 Machine Learning in Collision Risk Assessment

One promising strategy to alleviate collision hazards entails the utilization of ML instruments. Empirical datasets, structured as communications encompassing details regarding collision occurrences and near-miss events, establish a basis for the implementation of ML methodologies aimed at forecasting and averting collisions. For instance, the open-source Python library Kessler, as presented in [8], employs Bayesian neural networks to effectively forecast the progression of conjunction events. This instrument signifies a noteworthy progression within the domain, proffering a data-centric approach for evaluating collision risks.

The ESA has also made significant contributions to this domain. In [9], the ESA orchestrated an ML competition centered around an extensive and meticulously curated dataset. The objective of the competition was to motivate participants to devise optimal models for the evaluation of collision risk. The findings underscored both the difficulties associated with the formulation of a comprehensive training set and the prospective advantages of ML methodologies in tackling space collision challenges. These initiatives exemplify the increasing significance of ML in enhancing our capacity to forecast and avert collisions in LEO.

7.3.1.2 Challenges and Future Directions

Notwithstanding the advancements achieved, numerous obstacles persist. The inherently dynamic characteristics of LEO, coupled with the substantial quantity of objects present, obfuscate the formulation of models that possess universal applicability. Moreover, the accuracy of collision risk evaluations is contingent upon the quality and comprehensiveness of the data sets employed for the training of ML algorithms. Subsequent inquiries ought to prioritize the enhancement of data acquisition methodologies and the refinement of ML models to more effectively address the intricacies of the extraterrestrial environment.

The escalating risk of collisions in LEO necessitates the formulation of innovative strategies, with ML emerging as a formidable instrument for the prevention of such collisions. By leveraging empirical data sets and sophisticated algorithms, as elucidated in [8] and evaluated in [9], space agencies can significantly augment their capacity to forecast and alleviate collision hazards. Ongoing collaboration and advancements in technology will be critical to guarantee the sustainable utilization of LEO for forthcoming extraterrestrial endeavors.

7.3.2 Debris Identification and Removal Planning

The proliferation of orbital debris in LEO presents a considerable risk to both contemporary and prospective space exploration initiatives. Notwithstanding the execution of collision mitigation techniques, LEO remains inundated with debris resulting from satellite disintegration, collisions, and launch activities. As Wyler, the founder of OneWeb, astutely observed, the objective should be to "connect the world" rather than to contribute to the proliferation of "orbital refuse." To guarantee the enduring utilization of LEO, it is crucial to implement a bifocal strategy: preventing the emergence of novel debris and systematically eliminating pre-existing debris. In the subsequent sections, we will examine the technological and cooperative methodologies essential for addressing this urgent concern.

7.3.2.1 The Challenge of Space Debris Detection

The initial measure to alleviate the hazards associated with space debris involves its identification. Nevertheless, this undertaking is replete with obstacles. Space debris frequently exhibits a lack of discernible visual features, manifesting as amorphous aggregates with minimal reflectivity. Furthermore, the dynamic and tumultuous extraterrestrial milieu, exacerbated by the congested celestial panorama, culminates in an exceedingly low SNR. These elements render conventional detection methodologies inadequate for the dependable identification of debris.

Recent advancements in DL have indicated substantial potential in addressing these challenges. Scholars have engineered neural networks that are meticulously crafted for the identification of extraterrestrial debris. For instance, the CNN delineated in [10] employs local contrast maps that are extracted from space-based surveillance footage. This methodology capitalizes on spatiotemporal saliency data to augment detection robustness against time-varying noisy backgrounds. To further enhance detection throughput, the authors of [11] proposed a technique that segments acquired space imagery into diminutive, uniformly sized tiles, each annotated with a binary indicator denoting the presence of debris. These DL-oriented methodologies signify a noteworthy advancement in the precise and effective identification of space debris.

7.3.2.2 Active Debris Removal: Technological Solutions

Once debris has been identified, the subsequent challenge pertains to its elimination. RL has surfaced as an instrumental mechanism for orchestrating active debris removal operations, particularly within the framework of LEO satellite communication infrastructures. As evidenced by empirical studies utilizing the Iridium 33 system, RL functions as a proficient online reactive planner for missions aimed at the removal of multiple debris objects [12]. Nonetheless, it is important to note that not all debris can be swiftly de-orbited, particularly when it jeopardizes inter-satellite laser communication links. In response to this challenge, researchers have formulated a model for assessing interruption risks by employing ML methodologies. This model, as delineated in [13], establishes a basis for adaptive routing strategies that effectively mitigate the likelihood of disruptions in laser link communication during debris removal endeavors.

In addition to RL and ML, various alternative technological methodologies have been proposed for the removal of space debris. Robotic appendages, for instance, present a viable approach for the retrieval and de-orbiting of malfunctioning or non-functional spacecraft, thereby ensuring their incineration upon re-entry into the Earth's atmosphere. Although these methodologies exhibit considerable promise, their practical applicability necessitates thorough evaluation owing to the dynamic and uncertain characteristics of the space environment. This requirement engenders substantial computational demands for target identification, trajectory optimization, and monitoring [14].

7.3.2.3 Collaborative Efforts and Preventive Measures

Addressing the issue of space debris necessitates a concerted international collaboration. Nations and research institutions are required to engage in cooperative efforts to create a collective resource platform and formulate standardized protocols for debris mitigation. Simultaneously, proactive measures are of equal importance. In the phase of satellite deployment, the engineering of spacecraft ought to emphasize reusability and recyclability, as illustrated by SpaceX's Falcon series rockets [15]. Upon the conclusion of a satellite's operational lifespan, it is imperative that it be safely de-orbited to guarantee complete incineration in the Earth's atmosphere, thus averting its potential transformation into future debris.

For the purpose of addressing existing space debris, it is paramount to implement efficacious removal techniques, such as robotic arms, nets, harpoons, and oversized balloons, to ensnare and tow substantial debris back into the atmosphere for subsequent incineration. Furthermore, SSA systems, encompassing terrestrial radar and optical telescopes are integral to the detection and monitoring of debris throughout a satellite's operational lifespan. In instances where a potential collision is identified, satellites possess the capability to execute maneuvers to circumvent debris. The incorporation of SBRs or SBCs significantly augments the efficacy of collision avoidance strategies. The integration of AI tools to scrutinize data derived from SBRs and SBCs can also enhance the precision of detection and tracking, thus mitigating collision risks.

The escalating peril of orbital debris necessitates a comprehensive strategy that amalgamates sophisticated technological innovations with international collaboration. Through the application of AI, ML, and DL methodologies, we can significantly improve our capabilities in debris detection and removal. Concurrently, proactive measures and global partnerships are imperative to reduce the creation of additional debris and to uphold the sustainable utilization of LEO. Only through a concerted endeavor can we proficiently tackle the challenges presented by space debris and safeguard the prospects of space exploration.

7.4 Forward Error Control

7.4.1 Definitions

The conventional framework of a wireless communication system is comprised of a channel encoder and a channel decoder. The digital sequence from the source is encoded in the channel encoder and then sent through the channel. As the channel encoder uses error-correcting codes, the channel decoder employs the structure of the codes to correct the errors introduced in the transmission, thereby ensuring that the destination receives a sequence that is more reliably reconstructed.

FEC within the domain of wireless communications entails the incorporation of superfluous information to the transmitted data, thereby enabling the recipient to identify and rectify errors autonomously without necessitating retransmission, in contrast to the backward error correction methodologies employed in the data link layer that depends on retransmission protocols. The function of FEC at the receiving terminal is of paramount importance, as it discerns symbol inaccuracies within the received signals and, through established algorithms, accurately identifies the precise locations of these errors and rectifies them seamlessly. This mechanism is also known as automatic error correction, and it constitutes an effective strategy for optimizing channel utilization. In the realm of information theory, channel capacity theory is concerned with analyzing the boundaries of FEC performance for a designated channel. Concurrently, error correction coding theory is devoted to examining the particular encoding and decoding techniques associated with error correction codes (ECCs) utilized for FEC applications.

According to conventional error correction coding theory, for a code to exhibit error correction capabilities, it is imperative to append additional symbols to the original message. This is done to increase the disparity between codewords, known as the code distance. It is commonly held that these extra symbols added during transmission for error correction purposes lead to an expansion in the occupied bandwidth and a consequent reduction in the spectrum efficiency. Nevertheless, the integrated theory of coded modulation takes into account both modulation and error-correction coding simultaneously. By broadening the set of modulated signals, it becomes feasible to enhance the error correction capabilities of the coded modulation system without necessitating an augmentation of the channel bandwidth required. FEC has been extensively utilized in domains such as satellite communications, mobile communications, and high-bandwidth data transmissions.

7.4.2 FEC in Satellite Communications

7.4.2.1 Review of FEC Methods

The utilization of channel coding within the aerospace sector has its roots in facilitating deep space exploration initiatives. Among the SCSs that are publicly accessible, the Reed–Muller (RM) code and convolutional code represent the earliest adopted methodologies for channel coding. The RM code constitutes a block code that was introduced by Reed and Muller in 1954. In 1969, NASA implemented the first-order RM code, which featured a length of 32 bits, in the Mariner 6 spacecraft for the purpose of Mars exploration; this code comprised 6 information bits and 26 parity bits, with the receiver employing maximum likelihood decoding. In comparison to the uncoded system, it accomplished an encoding gain of approximately 3.2 dB at a BER of 10^{-5} [16].

Convolutional coding constitutes a distinct category of channel coding, initially introduced by Elias and colleagues in 1955. In contrast to block codes, this methodology does not encode individual segments subsequent to partitioning the information into discrete groups. Rather, it facilitates the continuous encoding of both input and output data streams. The encoded bits produced at each instance are intrinsically linked not only to the information bits input at that particular moment but also to inputs from preceding temporal intervals. In 1968, NASA successfully deployed the Pioneer 9 spacecraft, which functioned within a heliocentric trajectory. Between the years 1968 and 1990, this probe amassed data concerning the electromagnetic and plasma properties of interplanetary space while also conducting an extensive array of long-range communication experiments. The communication architecture employed in this probe utilized a systematic convolutional code characterized by a code rate of 1/2 and a memory depth of 20, as designed by [17]. The receiving apparatus implemented soft iterative decoding predicated on the Fano algorithm [18], enabling a remarkable encoding gain of approximately 6.9 dB. In the early 1970s, NASA systematically launched the Pioneer 10 and Pioneer 11 probes with the objective of investigating Jupiter, Saturn, and the outer regions of the solar system. These probes incorporated the convolutional code developed by Massey and Costello, which possessed a code rate of 1/2 and a constraint length of 32. In 1977, the prominent outer solar system missions, Voyager 1 and Voyager 2, employed a (2, 1, 7) convolutional code featuring a constraint length of 7 and utilized Viterbi decoding, achieving an encoding gain of 5.1 dB. This code was subsequently endorsed as the standard interstellar transmission code by the CCSDS and has seen widespread application. Following this, Earth-orbiting satellite systems, such as Globalstar and Iridium, also implemented convolutional codes with code rates of 1/2 and 3/4, respectively.

In the year 1960, Reed and Solomon introduced the Reed–Solomon (RS) code. Nevertheless, owing to the absence of a straightforward soft iterative decoding technique during that era, its application in satellite communication was not realized initially. In 1967, Forney put forth the notion of a concatenated code. Employing the RS code as the outer code in this concatenated scheme could facilitate remarkable error correction capabilities. The Mariner 9 Mars orbiter, which was launched in 1971, incorporated a concatenated code architecture, utilizing the (6, 4) RS code as the outer code and the (32, 6) RM code as the inner code. Although this coding structure exhibited a relatively low code rate, its superior error detection and correction capabilities significantly contributed to the resounding success of the Mariner 9 Mars orbiter. It effectively transmitted a total of 7,329 photographs and mapped 85% of Mars at a resolution of 1–2 km. When the inner code employs a convolutional code in conjunction with Viterbi decoding, while the outer code utilizes an RS code, outstanding performance can be attained. This methodology has emerged as the prevailing standard in subsequent satellite communications. Both Voyager 1 and Voyager 2 similarly implemented concatenated codes with the objective of ensuring the reliability of data transmission in compressed form while traversing the outer reaches of the solar system, including Uranus and Neptune. Their inner code comprised the (2, 1, 7) convolutional code, whereas the outer code consisted of the (255, 223) RS code. In comparison to the non-concatenated convolutional code, an additional gain of 2 dB could be realized. Digital video broadcasting (DVB) likewise integrated a comparable concatenated code within its DVB-S standard. Its outer code was represented by the (204, 188) RS code, while the inner code was delineated by the (2, 1, 7) convolutional code. Code rates such as 2/3, 3/4, 5/6, and 7/8 could be attained following puncturing processing. Since the 1970s, the concatenated code combining the RS code and the convolutional code has been extensively employed in various satellites, including the Intelsat IS-7 to IS-9 series and the Eutelsat Hotbird series satellites, among others.

In 1993, Berrou et al. introduced the Turbo code, a specific type of coding that employs the parallel concatenation of several component codes, wherein the component codes are categorized as systematic recursive convolutional codes [19]. Owing to the implementation of the interleaved concatenation architecture, its performance has been enhanced to a significant degree in comparison to traditional convolutional codes. Nevertheless, the decoding complexity associated with the Turbo code remains comparatively elevated, particularly in scenarios involving lengthy codes. Given that its performance approaches the Shannon limit, it has been incorporated into various technical standards, including CCSDS, DVB-RCS, and IEEE 802.16. The inaugural commercial system that facilitates the implementation of Turbo codes in satellite communication is Inmarsat-3, through which users are capable of attaining a communication rate of 64 kbps via the satellite [19]. The iPSTAR satellite employs Turbo product codes alongside advanced modulation techniques. The code rate is adjustable within the range of 0.879–0.325, and it furnishes a transponder with a peak power output of 120 W, thereby enabling the reverse link to achieve a maximum throughput of 4.1 Mbps [20]. Furthermore, the application of Turbo codes has also been extended to satellite systems such as Anik F2, DirecTV, and DISH Network.

The LDPC code represents a specific category of block code that was initially introduced by Gallager during the 1960s. Nevertheless, the significantly limited computational capabilities prevalent during that period led to a general perception of this code as excessively intricate, resulting in suboptimal developmental progress. In 1996, Mackay et al. re-identified the LDPC code, positing that this coding mechanism possessed a highly effective structural design and could facilitate parallel decoding processes [21]. Following this, through ongoing enhancements to the LDPC code architecture, the performance metrics of irregular LDPC codes have surpassed those of Turbo codes, notably exhibiting the absence of an error floor. Consequently, the DVB-S2 technical standard has incorporated a concatenated coding architecture wherein the LDPC code functions as the inner code and the Bose–Chaudhuri–Hocquenghem (BCH) code serves as the outer code. The LDPC code length is established at 64,800 for the long code and 16,200 for the short code, with the code rate being selectable within the range of 1/4–8/9 [22]. In the most recent iteration of DVB-S2X, an intermediate code length of 30,780 has been introduced, accompanied by an expanded array of code rate options. Since the inception of DVB-S2 over the preceding decade, numerous satellite organizations have either adopted or devised analogous technical standards. Satellites operated by various entities, including Intelsat, Eutelsat, Telstar, and Viasat, predominantly utilize the concatenated coding framework advocated within DVB-S2. Simultaneously, the LDPC code is recognized as a principal alternative for channel coding technology within the 5G NR framework. The Technical Specification TS38.212 delineates the application domains and coding structure of LDPC codes within the 5G system [23].

7.4.2.2 Future Trends and Development

Following the extensive implementation of capacity-approaching codes exemplified by LDPC codes, it becomes imperative to investigate capacity-achieving codes that can ultimately attain the Shannon limit. Polar codes, pioneered by

Erdal Arikan in 2009, have emerged as a revolutionary coding paradigm owing to their distinctive characteristics and advantages over previous coding methodologies, including LDPC and Turbo codes. A principal advantage of polar codes is their demonstrable capacity-achieving efficacy under successive cancellation (SC) decoding, indicating their theoretical capability to realize the Shannon limit for symmetric binary-input memoryless channels. This attribute renders them highly effective for control channels, wherein reliability and minimal complexity are of paramount importance. In contrast to LDPC codes, which depend on iterative decoding and may experience error floors, polar codes demonstrate error-floor-free behavior, thereby guaranteeing consistent reliability even at diminished error rates. Moreover, polar codes exhibit reduced decoding latency when juxtaposed with Turbo codes, rendering them particularly apt for URLLC scenarios. Although LDPC codes are particularly proficient in high-throughput data channels due to their parallelizable architecture, polar codes prove to be more efficient for short to moderate block lengths, which are prevalent in control signaling. Furthermore, polar codes are mathematically sophisticated and comparatively easier to construct, providing a systematic methodology for code design. These merits have facilitated their integration into 5G for control channels, signifying a noteworthy transition from the preeminence of Turbo and LDPC codes in preceding generations of wireless communication.

The preceding years have observed a swift advancement in the field of polar coding. In 2009, Arikan proposed the theory of channel polarization and presented the sequence construction algorithm—the Monte Carlo algorithm under the binary error channel (BEC) condition, as well as the SC and belief propagation (BP) decoding algorithms [24]. Consequently, Mori proposed to construct polar codes using the density evolution algorithm, which is applicable to all channels [25]. In 2011, Arikan proposed the systematic code form of polar codes [26]. In 2012, Trifonov proposed to construct polar codes using Gaussian approximation, and this method is applicable to Gaussian channels [27]. When the length of polar codes is medium to short or short, their performance is not optimal. In 2015, Xu proposed an enhanced BP decoding algorithm [28]. Later in 2017, He proposed the construction of polar codes using polarization weights [29].

Polar codes demonstrate considerable potential for SCSs, attributable to their capacity-achieving efficacy and minimal decoding latency, factors that are paramount for dependable data transmission across extensive distances. Their proficiency in functioning effectively at shorter to moderate block lengths renders them particularly advantageous for control signaling and low-latency applications, such as real-time telemetry and command systems utilized in satellite operations. Furthermore, the performance of polar codes is characterized by an absence of error floors, thereby ensuring consistent reliability, even under low SNR conditions frequently encountered in satellite communications. Nonetheless, obstacles persist in tailoring polar codes to meet the distinctive requirements of satellite communications. The elevated computational complexity associated with SC list decoding, which is essential for attaining performance close to the theoretical optimum, may impose limitations on the constrained processing capabilities present in satellite platforms. Additionally, the prolonged propagation delays and variable channel conditions inherent to satellite environments necessitate robust adaptation mechanisms, which remain in the developmental stages for polar codes. Although polar codes have exhibited the potential to enhance spectral efficiency and reliability, further scholarly inquiry is imperative to refine their implementation for scenarios specific to satellite communications, such as addressing Doppler shifts and alleviating atmospheric disturbances. Upon the resolution of these challenges, polar codes could evolve into a revolutionary technology for next-generation satellite communication infrastructures.

7.5 Interference Coordination

7.5.1 Principles of Interference Coordination

7.5.1.1 Background

In the domain of wireless communication, the coordination of interference represents a paramount and exceedingly complex endeavor. Its inception can be linked to the phase of extensive implementation of cellular networks. As the geographical reach of cellular networks continues to broaden, the optimal utilization of finite wireless resources has emerged as a fundamental concern that necessitates immediate resolution. Within the cellular network architecture, a multitude of base stations and mobile devices concurrently exploit limited spectral resources, leading to frequent interference among signals emanating from disparate cells, which profoundly undermines the quality of communication and the overall efficacy of the system. For instance, in densely populated urban locales, where base stations are closely spaced, the interference between neighboring cells results in erratic signal strength, consequently causing a marked decline in data transmission rates [30]. In this milieu, interference coordination technology has been developed. Its principal objective is to mitigate the interference between cells and to enhance the efficiency of wireless resource utilization through judicious resource allocation and advanced signal processing techniques.

Turning to the field of satellite communication, the coordination of interference encounters a multitude of distinct challenges:

Wide-area coverage and long propagation delay: Satellite communication exhibits distinct characteristics. For GEO satellites, the scope of their signal coverage is remarkably broad, encompassing nearly one-third of the Earth's surface. Nevertheless, this extensive coverage attribute simultaneously engenders the issue of prolonged propagation delay. The propagation delay associated with GEO satellites typically approximates 250 ms. Such a significant delay presents a formidable challenge to dynamic coordination within SCSs. In scenarios requiring real-time communication, such as satellite telephony, the considerable delay is perceptible to both parties involved in the conversation, thereby detrimentally impacting the overall call experience. Furthermore, in the context of satellite data transmission, the extended delay also diminishes the efficiency of data transfer.

Spectrum competition and the co-existence of satellite and terrestrial systems: The frequency bands allocated for satellite communications, including the C-band and Ka-band, exhibit overlaps with the frequency bands designated for terrestrial 5G networks. This phenomenon engenders a substantial potential for mutual interference between SCSs and terrestrial 5G communication systems. The simultaneous transmission of satellite and terrestrial 5G signals within the same frequency band inevitably results in signal conflicts, consequently diminishing the overall quality of communication. Thus, to facilitate the harmonious co-existence of satellite and terrestrial systems while addressing the issue of mutual interference, it is imperative to prioritize the allocation and utilization of spectrum resources.

Rise of LEO mega-constellations: The LEO constellations epitomized by Starlink encompass a multitude of satellites, thereby constituting an extensive satellite network. Within such a network, the risks associated with self-interference among ISLs and user connections escalate markedly [31]. The transmission of signals among a plethora of satellites is prone to mutual interference, which detrimentally influences the quality of communication both among satellites and between satellites and end-users. For instance, the signals emitted by proximate satellites may coincide and interfere during the transmission phase, leading to a heightened BER of the signals.

Malicious interference threats: In the context of military operations, satellite communication assumes a pivotal role; however, it is concurrently subjected to the risks of adversarial interference or assaults associated with electronic warfare. Adversarial entities may emit disruptive interference signals aimed at compromising the integrity of the satellite communication link, thereby incapacitating the satellite's operational capacity. Consequently, it is imperative for SCSs to enhance their resilience and employ interference mitigation strategies to counteract malicious disruptions and guarantee the security and stability of communication channels.

7.5.1.2 Basic Principles

It is not arduous to discern from the intricate context delineated above that the fundamental concept of interference coordination is to mitigate the adverse effects of extraneous signals on the intended receiver via the optimization of resource allocation. Its essence is to find a balance between the capacity, fairness, and complexity of the system. The specific principles can be analyzed from the following dimensions:

Mathematical representation of interference: Mathematically, interference is usually quantified by the SINR. The definition of SINR is the ratio of the target signal power to the sum of the interference signal power and the noise power

$$\mathrm{SINR} = \frac{P_s}{P_i + N}, \tag{7.14}$$

where P_s represents the target signal power, P_i represents the interference signal power, and N represents the noise power. Generally, the larger the SINR value, the smaller the interference the target signal receives, and the better the communication quality. Through in-depth analysis and accurate calculation of SINR, the degree of interference can be evaluated, providing a strong basis for the formulation of subsequent interference coordination strategies.

1. **Resource Dimension Division**

   Frequency domain: In the frequency domain, the available spectrum resources are partitioned into distinct frequency bands. Various users or communication links may utilize different frequency bands for communication, thereby mitigating interference. Analogous to cellular networks, the technology of frequency division multiple access (FDMA) is employed, segmenting the entire frequency band into several sub-frequency bands, with each user being assigned a particular sub-frequency band for their communication needs.

   Time domain: In the time domain, utilizing the time division multiple access (TDMA) technology, temporal intervals are partitioned into discrete time slots. Distinct users engage in communication within these separate time slots. This methodology effectively mitigates the interference that may arise from simultaneous signals transmitted by multiple users. For instance, in certain satellite communication frameworks, various ground stations interface with the satellite during individual time slots.

   Space domain: By leveraging the inherent properties of the space domain and employing advanced smart antenna technologies such as multi-antenna arrays, it becomes feasible to generate directional beams that effectively focus the signal energy towards the intended user. Concurrently, through the application of null steering technology, the antenna radiation pattern is meticulously modified to establish a null in the direction of interference, thereby mitigating interference in alternative directions.

   Power domain: Through the modulation of transmission power, while ensuring the fulfillment of communication prerequisites, the objective is to minimize the interference experienced by other users to the greatest extent feasible. For instance, in certain distributed communication systems, each node adaptively modifies its transmission power in accordance with the proximity to the target node and the prevailing state of the communication channel.

Code domain: The implementation of CDMA technology enables distinct users to employ unique code sequences for the purpose of encoding information. At the reception point, the signals originating from various users can be effectively distinguished through the process of correlation demodulation, thereby facilitating the concurrent utilization of the same spectral and temporal resources by multiple users while concurrently minimizing interference among them.

2. **Core Methodologies**

Static coordination: Static coordination delineates the resource allocation framework during the system design phase and remains invariant with respect to temporal progression and variations in channel states. For instance, predetermined frequency bands, time slots, or code sequences are allocated in advance to various cells or satellites. This approach is comparatively straightforward to execute; however, it is characterized by a lack of adaptability and poses challenges in accommodating the dynamically evolving communication landscape.

Dynamic coordination: In contrast to static coordination, dynamic coordination entails the real-time adaptation of the resource allocation framework in response to instantaneous CSI, user demands, and interference variables. For instance, through the continuous monitoring of the SINR metric, if significant interference is detected in a specific area, resources such as frequency bands, time slots, or user transmission power levels in that locale are recalibrated dynamically. This approach facilitates a more effective adjustment to the evolving communication landscape; however, it imposes stringent requirements concerning the real-time operational efficacy and computational capabilities of the system.

Cooperative coordination: Cooperative coordination underscores the significance of multiple communication nodes collaboratively executing interference coordination through the mechanisms of information exchange and mutual cooperation. Utilizing the satellite communication network as a case study, numerous satellites are capable of exchanging CSI and interference data amongst themselves, thereby jointly optimizing the resource allocation of ISLs and user connections to attain a more effective interference coordination outcome. This approach facilitates the comprehensive utilization of the resources and information provided by multiple nodes; however, it necessitates the establishment of a robust cooperation framework and a reliable channel for information exchange.

7.5.2 Key Technical Principles

In accordance with the aforementioned foundational principles, a comprehensive set of fundamental technical tenets has been formulated, thereby facilitating the advancement and utilization of interference coordination technology.

7.5.2.1 Frequency-domain Coordination

Frequency-domain coordination partitions the spectral range into a central frequency band and an edge frequency band. The central frequency band is characterized by a high reuse factor, whereby users situated in the central region experience relatively diminished interference, thus allowing central users to fully utilize the entire frequency band and optimize the utilization of spectral resources. Conversely, the edge frequency band employs a low reuse factor, wherein edge users are restricted to utilizing only the non-overlapping segments of the spectrum, thereby effectively mitigating interference among edge users from adjacent cells.

In a practical satellite communication framework, the implementation of frequency-domain coordination technology can be realized through methodologies such as filter-bank multi-carrier (FBMC). The FBMC methodology facilitates the partitioning of a broadband signal into several narrow-band subcarriers, wherein each subcarrier may be independently modulated and transmitted. Through meticulous filter design, the spectral overlap between subcarriers is effectively minimized, consequently enhancing the rate of spectrum utilization. Concurrently, for users positioned at the edges, specific non-overlapping subcarriers are allocated via an intelligent algorithm to mitigate interference with neighboring cells. For instance, within a satellite communication network encompassing multiple regions, edge users situated in various regions can be assigned distinct non-overlapping sub-carrier sets to ensure that the communication quality across each region is maintained, even under constrained spectrum resources.

7.5.2.2 Space-domain Coordination

The coordination within the spatial domain predominantly depends on the utilization of multi-antenna array technology. By employing the multi-antenna array, a directional beam can be generated to focalize signal energy towards the intended user, thereby enhancing both the efficiency and intensity of the signal transmission. Concurrently, the application of null-steering technology facilitates the modulation of the antenna's radiation pattern to create a null in the direction of interference, effectively mitigating disturbances emanating from other directions.

In satellite communication ground stations, the utilization of space-domain coordination technology is of paramount significance. For instance, the implementation of PA antenna technology is employed. Through the manipulation of both the phase and amplitude of individual antenna elements, rapid beam scanning and directional adjustments can be realized. Upon the detection of an interference source, the PA antenna is capable of swiftly altering the beam orientation to establish a null in the direction of the interference, thereby effectively attenuating the interference signal [32]. Furthermore, it can be synergistically integrated with adaptive antenna technology. In accordance with real-time CSI and prevailing interference conditions, the weights assigned to the antenna are dynamically modified to further enhance beam formation and the positioning of the null, thereby augmenting the adaptability of the SCS to the intricate electromagnetic environment.

7.5.2.3 Dynamic Spectrum Sharing in Satellite Communication

Dynamic spectrum sharing in satellite communication is facilitated through the implementation of cognitive radio technology. This advanced technology possesses the capability to detect the spectrum occupancy status of terrestrial systems, enabling the SCS to dynamically allocate frequency bands for communication according to this data, thereby mitigating interference between satellite and terrestrial systems.

In the framework of the spectrum coordination mechanism delineated by the 3GPP NTN standard, the SCS employs a spectrum sensing module to conduct real-time surveillance of the spectrum utilization by the terrestrial 5G system [32]. Upon the identification of a specific frequency band that remains unutilized by the terrestrial 5G system, the SCS adeptly modifies its communication parameters via a spectrum decision algorithm, thereby facilitating access to this frequency band for communication purposes. Concurrently, to ensure the seamless integration of satellite-terrestrial communication, the SCS engages in information exchange with the terrestrial 5G system, disseminating the spectrum utilization strategy and communication parameters in order to mitigate the risk of potential interference. For instance, in metropolitan regions characterized by high demand for the terrestrial 5G network, the SCS strategically circumvents the frequently employed 5G frequency bands, opting instead for alternative idle frequency bands for communication; conversely, in remote areas marked by inadequate terrestrial 5G coverage, the SCS can exploit the spectrum resources with greater flexibility to enhance communication efficiency.

7.5.3 Application Scenarios of Interference Coordination in Satellite Secure Communication

7.5.3.1 Military Secure Communication

In the domain of military operations, the dependability and longevity of communication systems are of paramount significance. Geostationary satellites, characterized by their extensive coverage and comparatively stable orbital placements serve as essential resilient backup links for the command center. In scenarios where adversarial interference or other communication link disruptions occur, the implementation of interference coordination technology is of critical importance.

During periods of armed conflict, adversarial forces may implement a range of disruptive measures, including broad-spectrum jamming and precision-targeted narrow-band jamming, in an effort to undermine the integrity of satellite communication networks. In response, the GEO satellite communication framework can employ frequency-hopping methodologies to rapidly transition among a multitude of frequencies, thereby circumventing the designated interference bands utilized by adversaries. Concurrently, advanced error-correction coding techniques, such as LDPC codes, are deployed to encode the transmitted information, thereby enhancing the resilience of the data against interference. Furthermore, intelligent antenna technology is leveraged to precisely orient the communication beam toward the military unit requiring connectivity, thereby minimizing signal leakage in alternative directions and mitigating the likelihood of interception and disruption by hostile entities. Through the integrative application of these interference mitigation technologies, the security of communications between the command center and frontline personnel is safeguarded, facilitating the accurate dissemination of military directives.

7.5.3.2 Anti-congestion Interference in Commercial Constellations

The LEO constellations, as delineated by Starlink, consist of thousands of satellites that collectively establish an expansive and intricate communication network. Within this densely populated satellite framework, issues related to self-interference and congestion are markedly pronounced.

When the volume of satellite communication traffic within a specific region escalates, leading to congestion within the frequency bands, the satellite is capable of autonomously detecting the utilization of the surrounding spectrum via spectrum sensing technology. By employing machine-learning algorithms, both historical spectrum data and contemporary service requirements are meticulously analyzed to forecast the trends in spectrum variations, enabling the dynamic adjustment of its communication frequency bands, thereby mitigating interference sources, and selecting relatively underutilized frequency bands for effective communication. Concurrently, with regard to ISLs, resource allocation algorithms are optimized based on factors such as the inter-satellite distance, relative motion velocities, and communication demands, facilitating the judicious arrangement of communication time slots and power levels among various satellites. For instance, a hybrid approach incorporating TDMA and CDMA is employed to assign distinct time slots and code sequences to different ISLs, thereby minimizing signal interference among satellites, guaranteeing the stability and efficacy of communication services within commercial constellations, and ensuring uninterrupted Internet access for users globally.

7.5.3.3 Satellite–terrestrial Converged Network

With the synergistic advancement of satellite communication and terrestrial communication, the establishment of a unified satellite-terrestrial network has emerged as a prevailing trend in the field of communication. Within the satellite-terrestrial converged network, it is imperative for satellites and terrestrial base stations to allocate and utilize spectrum resources collaboratively, which unavoidably results in the occurrence of mutual interference.

Interference coordination technology systematically orchestrates the utilization of the electromagnetic spectrum by satellites and terrestrial base stations through precise spectrum sensing methodologies and sophisticated resource allocation algorithms. Both SCSs and terrestrial base stations are outfitted with spectrum-sensing apparatuses to continuously monitor the surrounding electromagnetic milieu in real-time. Upon the identification of idle frequency bands by the SCS, it engages in information exchange with terrestrial base stations, subsequently ascertaining that no interference will ensue for terrestrial communications, thereby facilitating dynamic access to these frequency bands for transmission purposes. Concurrently, terrestrial base stations possess the capability to modulate their signal transmission power and coverage in accordance with the exigencies of satellite communication. For instance, in regions characterized by weak satellite signal coverage, terrestrial base stations may judiciously diminish their transmission power to mitigate interference with satellite signals; conversely, during intervals of reduced terrestrial communication demands, terrestrial base stations can proactively relinquish certain spectrum resources for satellite communication, thereby accomplishing a seamless integration and effective collaboration between satellite and terrestrial communication systems, ultimately delivering users an expansive and superior quality of communication services.

7.5.4 Advantages and Disadvantages

Every technological advancement possesses dual facets, and interference coordination technology is not an anomaly. While it presents numerous benefits, it concurrently encompasses certain drawbacks.

Advantages

1. **Improving spectrum efficiency and system capacity:** By implementing judicious interference management strategies, including the optimized distribution of resources across the frequency domain, time domain, spatial domain, and other parameters, it becomes feasible for a greater number of users to engage in communication within the constraints of limited spectral resources. This, in turn, results in improved spectral utilization efficiency and consequently bolsters the overall capacity of the system.

2. **Enhancing interference resilience:** The implementation of interference management methodologies, including null-steering techniques and adaptive resource distribution, has the potential to significantly mitigate interference signals, bolster the system's interference resistance, and elevate the quality of communication.
3. **Enhancing security and survivability:** In response to deleterious disruptions, interference coordination technology has the capacity to facilitate a seamless communication link through the adaptive modification of resource allocation and signal processing methodologies, thereby augmenting the security and resilience of the SCS.

Disadvantages

1. **High technical complexity:** The coordination of interference necessitates the allocation of resources across various dimensions alongside sophisticated signal processing algorithms. For instance, real-time monitoring of channel states, dynamic adjustments of resources within adaptive coordination frameworks, inter-node information exchange, and collaborative processing in cooperative coordination significantly augment the technical complexity inherent to the system.
2. **Propagation delay and synchronization problems:** In the realm of satellite communication, the inherent long propagation delay results in a lagging interaction of information and resource allocation in interference coordination, thereby complicating the synchronization of the overall system. For instance, in scenarios involving dynamic coordination, the aforementioned delay may hinder the timely adjustment of resources, subsequently impacting the efficacy of communication.
3. **Resource allocation conflicts:** Within a multi-user framework, diverse users manifest varying resource exigencies and priorities, which may culminate in conflicts regarding resource allocation. For instance, when numerous users simultaneously vie for constrained spectrum resources, the equitable and effective distribution of such resources emerges as a complex challenge.

References

1 Aaron D Wyner. The wire-tap channel. *Bell Syst. Tech. J.*, 54(8):1355–1387, 1975.
2 Hang Deng, Ziyun Fu, Xiaqing Miao, Shuai Wang, Gaofeng Pan, and Jianping An. Secure uplink transmissions in hybrid RF-UWOC space-ocean systems. *IEEE Trans. Wirel. Commun.*, 23(7):7816–7832, 2023.

3 Ali Abdi, Wing C Lau, Mohamed-Slim Alouini, and Mostafa Kaveh. A new simple model for land mobile satellite channels: First- and second-order statistics. *IEEE Trans. Wirel. Commun.*, 2(3):519–528, May 2003.

4 Bin Li, Zesong Fei, Caiqiu Zhou, and Yan Zhang. Physical-layer security in space information networks: A survey. *IEEE Internet Things J.*, 7(1):33–52, 2019.

5 Andrea Goldsmith. *Wireless Commun.* Cambridge University Press, 2005.

6 Il-Min Kim, Byoung-Hoon Kim, and Joon Kui Ahn. Ber-based physical layer security with finite codelength: Combining strong converse and error amplification. *IEEE Trans. Commun.*, 64(9):3844–3857, 2016.

7 Yury Polyanskiy, H Vincent Poor, and Sergio Verdu. Channel coding rate in the finite blocklength regime. *IEEE Trans. Inf. Theory*, 56(5):2307–2359, May 2010.

8 Giacomo Acciarini, Francesco Pinto, Francesca Letizia, et al. Kessler: A machine learning library for spacecraft collision avoidance. In *8th European Conference on Space Debris*, pages 1–9, Apr. 2021.

9 Thomas Uriot, Dario Izzo, Luís F Simões, et al. Spacecraft collision avoidance challenge: Design and results of a machine learning competition. *Astrodynamics*, 6(2):121–140, Apr. 2021.

10 Jiang Tao, Yunfeng Cao, Likui Zhuang, et al. Deep convolutional neural network based small space debris saliency detection. In *Proceeding of the 2019 25th International Conference on Automation and Computing*, pages 1–6, Lancaster, UK, 5–7 Sep. 2019.

11 Yaobing Xiang, Jiangbo Xi, Ming Cong, et al. Space debris detection with fast grid-based learning. In *Proceedings of the IEEE 3rd International Conference of Safe Production and Informatization*, pages 205–209, Chongqing City, China, 28–30 Nov. 2020.

12 Jianan Yang, Yu Hen Hu, Yong Liu, et al. On the application of reinforcement learning in multi-debris active removal mission planning. In *2019 IEEE 28th International Symposium on Industrial Electronics (ISIE)*, pages 605–610, Vancouver, BC, Canada, IEEE, 12–14 Jun. 2019.

13 Zhuangzhuang Ma, Yongli Zhao, Wei Wang, et al. Adaptive snapshot routing based on space debris risk perception in satellite optical networks. In *2021 International Conference on Optical Network Design and Modeling (ONDM)*, pages 1–6, Gothenburg, Sweden, IEEE, 28 Jun.–01 Jul. IEEE, 2021.

14 Wei Zhang, Feng Li, Junlin Li, et al. Review of on-orbit robotic arm active debris capture removal methods. *Aerospace*, 10(1):13, Dec. 2022.

15 Wei Zhang, Tao Wu, and Hong Ma. Discussion on the development direction of intelligent integrated space TT&C network. In *2021 7th International Conference on Computer and Communications (ICCC)*, pages 459–463, Chengdu, China, IEEE, 10–13 Dec. 2021.

16 Elwyn Berlekamp and Lloyd Welch. Weight distributions of the cosets of the (32, 6) reed-muller code. *IEEE Trans. Inf. Theory*, 18(1):203–207, 1972.

17 Shu Lin and H Lyne. Some results on binary convolutional code generators (corresp.). *IEEE Trans. Inf. Theory*, 13(1):134–139, 1967.

18 Robert Fano. A heuristic discussion of probabilistic decoding. *IEEE Trans. Inf. Theory*, 9(2): 64–74, 1963.

19 Claude Berrou, Alain Glavieux, and Punya Thitimajshima. Near Shannon limit error-correcting coding and decoding: Turbo-codes. 1. In *Proceedings of ICC'93-IEEE International Conference on Communications*, volume 2, pages 1064–1070. IEEE, 1993.

20 William Thesling, Mark Vanderaar, Mark Thompson, Peter Hamilton, Paiboon Panuwattanawong, and Richard Gedney. Two-way internet over iPSTAR using advanced error correction and dynamic links. In *20th AIAA International Communication Satellite Systems Conference and Exhibit*, page 1944, 2002.

21 David JC MacKay. *Information Theory, Inference and Learning Algorithms*. Cambridge University Press, 2003.

22 ETSI. Digital video broadcasting (DVB); Second generation framing structure, channel coding and modulation systems for broadcasting, interactive services, news gathering and other broadband satellite applications; Part 1: DVB-S2 V1.4.1. R, ETSI, 2014.

23 3GPP. 3GPP TS 38.212: Multiplexing and channel coding V16.0.0 (Release 16). R, 3GPP, 12 2019.

24 Erdal Arikan. Channel polarization: A method for constructing capacity-achieving codes for symmetric binary-input memoryless channels. *IEEE Trans. Inf. Theory*, 55(7):3051–3073, 2009.

25 Ryuhei Mori and Toshiyuki Tanaka. Performance of polar codes with the construction using density evolution. *IEEE Commun. Lett.*, 13(7): 519–521, 2009.

26 Erdal Arikan. Systematic polar coding. *IEEE Commun. Lett.*, 15(8):860–862, 2011.

27 Peter Trifonov. Efficient design and decoding of polar codes. *IEEE Trans. Commun.*, 60(11):3221–3227, 2012.

28 Jingwei Xu, Tiben Che, and Gwan Choi. Xj-bp: Express journey belief propagation decoding for polar codes. In *2015 IEEE Global Communications Conference (GLOBECOM)*, pages 1–6, 2015.

29 Gaoning He, Jean-Claude Belfiore, Ingmar Land, et al. Beta-expansion: A theoretical framework for fast and recursive construction of polar codes. In *GLOBECOM 2017–2017 IEEE Global Communications Conference*, pages 1–6, 2017.

30 Junyu Liu, Min Sheng, Lei Liu, and Jiandong Li. Interference management in ultra-dense networks: Challenges and approaches. *IEEE Netw.*, 31(6): 70–77, 2017.

31 Navid Heydarishahreza, Tao Han, and Nirwan Ansari. Spectrum sharing and interference management for 6G LEO satellite-terrestrial network integration. *IEEE Commun. Surv. Tutor.*, page 1, 2024.

32 Mingqian Liu, Zhaoxi Wen, Lei Jin, and Ming Li. Intelligent estimation of frequency domain parameters for satellite communication interference with alpha-stable noise. In *2023 IEEE 98th Vehicular Technology Conference (VTC2023-Fall)*, pages 1–5, 2023.

Part IV

Design Criteria for Security SCSs

8

Parameter Selections

The formulation of secure SCSs is intrinsically intricate, chiefly attributable to the necessity of reconciling various potentially discordant design parameters. In this segment, we shall delineate pragmatic design principles for LEO SCS from the vantage points of security and dependability. These principles are extracted from a comprehensive analysis of existing scholarly works and insights garnered from historical experiences, emphasizing the security and reliability prerequisites and the corresponding strategies to mitigate these challenges. By scrutinizing these matters meticulously, we aspire to provide more definitive and efficacious recommendations for forthcoming designs, thereby ensuring that SCSs can uphold elevated benchmarks of security and reliability in practical applications [1].

8.1 Orbits

Figure 8.1 presents a diagram showing key orbital parameters of satellites that define the characteristics of the Earth's orbits. These parameters include the following:

- Apogee: The point in the orbit that is farthest from Earth
- Perigee: The point in the orbit that is closest to Earth
- Line of perigee and apogee: The line passing through the Earth's center, connecting the apogee and perigee
- Ascending node: The point where the orbital plane intersects the equatorial plane as the satellite moves from south to north
- Descending node: The point where the orbital plane intersects the equatorial plane as the satellite moves from north to south
- Line of nodes: Line passing through the center of Earth connecting the ascending and descending nodes

Safeguarding Satellite Communications: Issues, Challenges, and Solutions, First Edition.
Jianping An, et al.

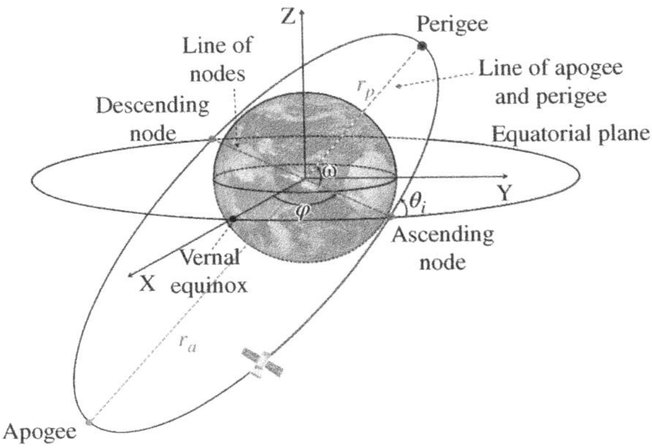

Figure 8.1 Earth-orbiting satellite parameters.

- Argument of perigee (ϖ): Angle within the orbital plane between the ascending node and the perigee
- Right ascension of the ascending node (φ): The angle within the equatorial plane, measured from the vernal equinox eastward to the ascending node

The eccentricity is a measure of the "roundness" of the orbit and is determined by

$$\eta = \frac{r_a - r_p}{r_a + r_p}, \tag{8.1}$$

where η is the orbital eccentricity. r_a and r_p stand for the distance from the Earth's center to the apogee and perigee, respectively. The greater the eccentricity, the more "flattened" the ellipse becomes. The circular orbit is a special case of an elliptical orbit, where the major and minor axes are equal (eccentricity equals 0), i.e., elliptical orbits $0 < \eta < 1$ and circular orbits $\eta = 0$. The inclination (θ_i) is the angle between the orbital plane and the Earth's equatorial plane. Orbits with a certain inclination are referred to as inclined orbits. Orbits within the equatorial plane ($\theta_i = 0°$) are called equatorial orbits [2]. An orbit with an inclination of 90° is called a PO. The orbit can be either elliptical or circular, depending on the satellite's velocity and direction when it enters orbit. Prograde orbits refer to those in which the satellite moves in the direction of Earth's rotation, with an inclination ranging from 0° to 90° [3]. In retrograde orbits, the satellite moves in the opposite direction to Earth's rotation, with inclination ranging from 90° to 180°. Most satellites are launched into prograde orbits because Earth's rotation speed helps increase the satellite's velocity in orbit, reducing the energy required for launch and insertion. There is a wide variety of combinations of orbital parameters. These parameters define a set of values that uniquely determine the position of a satellite in orbit [4].

The minimum number of parameters required is six, including (1) eccentricity; (2) half-length axis; (3) perigee moment; (4) ascending node declination; (5) inclination; and (6) perigee spoke angle [5]. The parameters delineated above facilitate the unequivocal determination of the absolute (i.e., inertial) coordinates of the satellite at any specified instant, thus enabling the elucidation of the satellite's trajectory and the forecasting of its prospective position. Orbital coordinates of satellites are articulated in terms of sidereal time as opposed to solar time. Solar time is fundamentally derived from one complete revolution of Earth in relation to Sun, serving as the foundation for all global temporal standards. Conversely, sidereal time is predicated upon a complete revolution of Earth relative to a fixed reference star. Given that sidereal time is contingent upon a full revolution of Earth with respect to a remote reference star at infinity rather than Sun, a mean sidereal day is approximately 0.3 percent shorter than a mean solar day.

8.1.1 Impact of Orbital Parameters on Security Communications

The determination of satellite orbits is of paramount importance in the formulation of secure and dependable satellite communication frameworks, as illustrated in Figure 8.2. Orbital parameters, particularly those pertaining to altitude and inclination, exert a substantial effect on both the security and operational efficacy of the satellite system. As the quantity of satellites in orbit escalates, especially

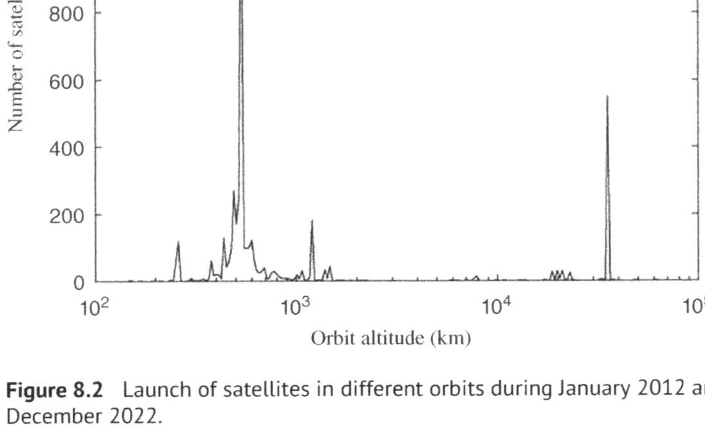

Figure 8.2 Launch of satellites in different orbits during January 2012 and December 2022.

with the emergence of LEO mega-constellations, the significance of meticulous and informed orbit selection becomes increasingly imperative [6]. This section will analyze how various orbital selections directly influence the security, safety, and reliability of communication systems, emphasizing how particular decisions can either enhance or hinder system performance [7]. In the realm of LEO satellites, the conventional altitude range is situated between 500 and 2,000 km; however, the burgeoning presence of mega-constellations renders these altitudes progressively congested [8]. A principal factor in the selection of orbital altitude is the satellite's coverage area. A reduction in latency enhances the speed and reliability of communication links, thereby rendering them less susceptible to interception or jamming.

At elevated altitudes, such as those encountered in MEO or GEO, the coverage area experiences a substantial augmentation, thereby potentially diminishing the requisite number of satellites for comprehensive coverage. Nevertheless, the escalation of distance from the terrestrial surface introduces an increased signal propagation delay and latency. In the context of security, this delay may complicate the execution of real-time cryptographic processes, resulting in diminished responsiveness during encryption and decryption activities. For communications of a highly sensitive nature, where minimal latency is paramount to safeguarding time-sensitive information, LEO continues to be the preferred alternative.

Within a constellation architecture, multiple satellites typically exhibit varying orbital inclinations. When these satellites collaborate, the potential for enhanced communication security is significantly realized. For instance, the avoidance of communication interference, or the fortification against such interference, can be achieved through the dynamic modulation of orbital inclination. The diversity of orbital inclinations contributes to system redundancy and enhances the overall security and resilience against potential attacks on the communication infrastructure. Communications requiring a high level of security necessitate satellites capable of rapidly modifying their orbits or operational actions under specific circumstances to evade prediction or interference from adversarial entities. The consideration of inclination adjustments during the orbital design phase can augment the mobility and evasive capabilities of satellites, thereby improving communication security. The inclination of a satellite's orbit dictates the temporal and angular parameters during which the satellite traverses various ground stations. An inadequately designed inclination may result in frequent transitions between ground stations, a process that could be exploited by malevolent actors to perpetrate attacks or expropriate communication data. This potential risk can be effectively mitigated through the strategic adjustment of the inclination angle.

The implementation of multi-orbit collaboration circumvents the vulnerabilities inherent in a singular orbital configuration [9]. For instance, the heightened susceptibility to interference posed by LEO satellites operating at lower altitudes can be counterbalanced by satellites positioned in higher orbits, thereby enhancing the redundancy of the entire system. Satellites occupying diverse orbits can leverage distinct fields of view and varying orbital mobility to distribute the communication load, thus bolstering the system's capacity to withstand interference and counterattacks. Multi-orbit collaboration affords increased steering space and flexibility, enabling LEO satellites to elude detection or attack trajectories through orbital modifications. The concept of satellite cluster design entails the systematic arrangement of multiple satellites in the same or disparate orbits, functioning in concert with meticulous orbit control and management of their relative positions. Satellite clusters can collectively mitigate risks through coordinated responses in the face of threats, particularly when confronted with sophisticated jamming or attacks, thereby diminishing single points of failure through the flexible adjustment of positions and task allocations [10]. By deploying multiple satellites across different orbits, satellite clusters can provide redundancy, ensuring that the operational integrity is maintained even in the event of individual satellite attacks.

Satellites are typically outfitted with a propulsion mechanism that facilitates the correction of their orbital trajectories. Through systematic orbital adjustments, satellites are able to mitigate deviations from their designated paths due to atmospheric drag and maintain their positions within a specified orbital envelope, thereby preventing disruptions in communication services. The satellite's attitude control system (ACS) actively seeks to achieve the optimal orientation, ensuring that there is no degradation of attitude that could compromise the integrity of the communication link. Advanced coding and modulation methodologies are employed to ensure that the communication signal retains a superior signal-to-noise ratio, thereby minimizing data loss even when subjected to external perturbations.

With the proliferation of LEO satellite constellations, the spatial distribution of satellites has become increasingly congested. Inadequate orbital design may precipitate collisions between satellites, which could not only result in the failure of the satellites involved but also generate debris that compromises the robustness and safety of communications for other operational satellites. Satellites in orbit are influenced by an array of external forces, including the nonuniform gravitational influence of Earth, solar radiation pressure, and solar wind, which may induce shifts in the satellite's orbit. The primary objective of orbital design is to ascertain an appropriate type and configuration of orbit. Based on the specific requirements of satellite applications, various orbital types may be selected. Utilizing orbital prediction models alongside collision avoidance algorithms, the relative motion among satellites can be monitored in real time, allowing for proactive orbital adjustments to be implemented. To enhance the robustness of communications, satellites are organized into formations, comprising groups of multiple satellites. These satellites efficiently transmit data via direct optical communication links (e.g., FSO) or traditional RF links. The relative spatial configurations of the satellites must be meticulously considered during the formation design process to ensure that they maintain communication within a stable line of sight while mitigating potential frequency interference.

For communications characterized by a high degree of reliability, the GEO is extensively employed in terrestrial communication infrastructures owing to its inherent stability [10]. Despite the inherent challenges associated with direct communications between satellites and terrestrial stations during deep-space exploration missions, GEO orbits can be strategically utilized as relay satellites to facilitate effective signal transmission with deep-space probes. For instance, in missions targeting the lunar surface or the exploration of Mars, GEO satellites can furnish continuous and stable communication support for probes that are situated at significant distances from Earth by functioning as intermediary relay stations.

In the engineering of satellites designated for lunar or Martian orbits, it is imperative to account for communication delays with Earth, as well as potential occlusion challenges that may arise. To uphold the ongoing reliability of communications, it is a common practice to design multiple satellites that collectively form a network. In the context of Mars missions, satellites are typically synchronized through a relay communications network interfacing with ground stations to mitigate the risks associated with the failure of any singular satellite that could detrimentally impact the mission [11]. As illustrated in Figure 8.3, ensuring the high reliability of deep-space communications may prove challenging for a single satellite to satisfy the continuous communication requirements of the mission, particularly during the signal transmission process, which may encounter scenarios such as blockage or loss of line of sight. Consequently, the establishment of a multi-satellite cooperative relay network is fundamental to guaranteeing the stability and reliability of deep-space communication links.

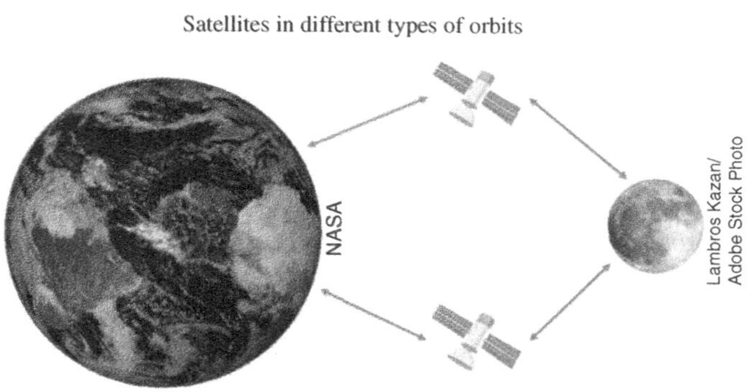

Figure 8.3 Deep space communications.

8.2 Frequencies and Waves

8.2.1 Frequency Selections

As a pivotal component of SCSs, the process of selecting frequencies tailored to various business applications and usage contexts facilitates adaptable, dependable, and effective service provision and transition. Typically, the challenge of frequency selection within satellite communications can be delineated into two primary categories: frequency selection for diverse service offerings and frequency selection for the mitigation of interference.

8.2.1.1 Frequency Selections for Heterogeneous Services

As is the case in the selection of orbital positions, the acquisition of frequency resources and the establishment of frequency allocation frameworks for satellite communications necessitate formal requests to the ITU. This procedural requirement for "prior authorization of use" ensures a structured approach to spectrum management; nevertheless, it concurrently demands a heightened level of meticulousness in the advanced design of frequency selection methodologies. Specifically, heterogeneous SCSs may exhibit varying degrees of exigencies concerning reliability and security, thereby necessitating frequency selection schemes that are tailored to the specific user demographics and application contexts.

For example, in the context of satellite communication applications within the IoT, the utilization of the L-band, characterized by minimal propagation losses, emerges as a suitable option. This assertion is particularly pertinent for IoT devices that possess constrained energy resources and exhibit relatively modest QoS requirements. Conversely, in relation to high-throughput backhaul services facilitated by satellites, the K-band is favored due to its capacity to support data links that achieve elevated data transfer rates.

As illustrated in Figure 5.1, satellite constellations that cater to relatively low data rate demands (such as Inmarsat, Globalstar, Iridium, and Kepler) predominantly opt for lower frequency bands (ranging from the L-band to the C-band), whereas constellations designed for Internet connectivity or high-throughput services (including Starlink, OneWeb, and Telesat) generally select higher frequency bands, extending from the Ku-band to the U-band. These higher frequency bands afford more available frequency resources, increased bandwidth, and diminished levels of interference.

In conclusion, it is imperative to meticulously design and allocate the most suitable frequency bands to distinct categories of users or services in order to effectively meet the diverse communication requirements inherent in satellite networks.

8.2.1.2 Frequency Selections for Interference Management

In the realm of satellite communications, the management of interference constitutes a vital component that guarantees both the reliability and efficacy of the system. Interference may emanate from a multitude of sources, encompassing CCI, adjacent channel interference, as well as interference from alternative satellite systems or terrestrial communication networks. The implementation of effective frequency selection is imperative to alleviate these sources of interference and to uphold the QoS for all users. This subsection examines the methodologies and strategies for frequency selection that are specifically designed to address and mitigate interference within SCSs.

1. **Multiple Access Techniques for Interference Management**

   One of the fundamental methodologies for managing interference is the implementation of multiple access techniques that facilitate the efficient utilization of the frequency spectrum among numerous users. Within the realm of satellite communications, a variety of multiple access techniques are utilized, encompassing FDMA, TDMA, CDMA, and more contemporaneously, orthogonal frequency division multiple access (OFDMA) and sparse code multiple access (SCMA).

   FDMA, for example, allocates distinct frequency bands to individual users, thereby mitigating CCI. Nevertheless, FDMA necessitates a meticulous frequency allocation strategy to circumvent interference between adjacent channels, particularly in circumstances where frequency resources are constrained. Conversely, TDMA segments the available time into discrete slots and assigns these slots to various users, which can be efficacious in interference management within temporally variable environments. CDMA employs spreading codes to differentiate among users, thereby facilitating multiple users to concurrently utilize the same frequency band and time slot. This methodology can be especially advantageous in high-interference environments, as it offers a degree of interference resilience through the deployment of unique codes.

   More sophisticated multiple-access techniques, such as OFDMA and SCMA, have garnered increased attention in recent years due to their capacity to enhance spectral efficiency and optimize interference management. OFDMA integrates FDMA and TDMA by partitioning the frequency band into orthogonal subcarriers, each of which may be allocated to different users. This strategy not only augments spectral efficiency but also enhances interference management by dynamically assigning subcarriers in accordance with the prevailing interference conditions. SCMA, conversely, is a sparse coding-based multiple-access technique that enables multiple users to concurrently share the same time–frequency resources through the utilization of sparse codes, thereby diminishing interference and enhancing spectral efficiency.

2. **Spectrum Sensing for Dynamic Frequency Selections**
In conjunction with various access methodologies, spectrum sensing assumes an essential function in the management of interference by facilitating adaptive frequency allocation. Spectrum sensing encompasses the surveillance of the frequency spectrum to ascertain the existence of interference or unoccupied frequency bands that may be leveraged for communication purposes. This aspect is particularly critical in satellite communications, wherein the frequency spectrum constitutes a limited resource and interference can markedly influence the efficacy of the system.
Cognitive radio, a pivotal facilitator of dynamic spectrum utilization, is profoundly reliant on spectrum sensing to discern and capitalize on underutilized frequency bands. Within the framework of satellite communications, cognitive radio methodologies can be employed to dynamically select frequencies that exhibit a reduced susceptibility to interference, thus enhancing the overall performance of the system. For instance, through real-time spectrum sensing, an SCS is capable of circumventing frequencies that are significantly affected by terrestrial networks or other satellite systems.
Spectrum sensing methodologies can be broadly classified into three principal categories: energy detection, matched filter detection, and cyclostationary feature detection. Energy detection represents the most straightforward and extensively utilized approach, wherein the energy within a designated frequency band is quantified to ascertain the presence of a signal. Conversely, matched filter detection entails the correlation of the received signal with a predetermined signal waveform to identify the existence of a specific signal. Cyclostationary feature detection capitalizes on the cyclostationary characteristics of signals, such as the periodicity inherent in the signal's spectrum, to detect the presence of a signal.
In the realm of satellite communications, the selection of a spectrum sensing technique is contingent upon the particular requirements of the system, including the requisite detection precision, complexity, and computational resources. For example, energy detection is computationally efficient; however, it may lack the accuracy of alternative techniques in environments characterized by elevated levels of noise or interference. Conversely, matched filter detection and cyclostationary feature detection provide enhanced detection accuracy, albeit at the expense of heightened computational complexity.
3. **Challenges and Future Directions**
Notwithstanding the advancements in multiple access methodologies and spectrum sensing technologies, the management of interference within SCSs continues to present formidable challenges. A principal obstacle is the inherently dynamic character of the interference milieu, which is subject to rapid fluctuations resulting from satellite mobility, the initiation of novel services,

or the emergence of interfering entities. In order to mitigate this issue, forthcoming research endeavors ought to prioritize the formulation of adaptive frequency selection methodologies capable of dynamically responding to the evolving interference landscape.

Additionally, another significant challenge arises from the surging demand for spectrum resources, which is propelled by the widespread adoption of satellite-based applications such as the IoT, high-capacity data services, and satellite-enabled 5G networks. This necessitates the innovation of more efficacious frequency allocation frameworks that can optimize the utilization of the available spectrum while concurrently minimizing interference.

In conclusion, the process of frequency selection for interference management constitutes a pivotal element of satellite communications that necessitates meticulous evaluation of multiple access methodologies and spectrum sensing techniques. By harnessing sophisticated multiple access strategies and dynamic spectrum sensing capabilities, satellite communication frameworks can realize enhanced interference management, thus guaranteeing reliable and efficient service provision in an increasingly intricate and congested spectrum environment.

8.2.2 Waveform Selections

8.2.2.1 Spread Spectrum Techniques

Due to the inherently open characteristics of satellite communication environments, it is crucial to safeguard transmitted messages and the corresponding transmission behaviors from eavesdropping and detection by noncooperative entities. Consequently, the selection of waveforms in satellite communications constitutes a critical subject that underpins the essential capabilities pertaining to reliability and security.

Considering the fundamental requirements for concealment and resilience against interference, SS waveforms emerge as the most viable option for satellite communications. The foundational principle of SS techniques involves the expansion of the signal's frequency spectrum through the application of a spreading code to either the time-domain signal sequence or the frequency-domain hopping pattern. Typically, the spreading codes are selected as PRN sequences, which can be designed to be sufficiently brief to facilitate efficient data transmission with relatively low latency, while being complex enough to hinder potential detection and decoding efforts. Based on the data elements influenced by the PRN sequences, SS methodologies can be classified into DSSS and FHSS. In general, SS techniques provide numerous advantages over non-spread-spectrum waveforms, as outlined in the following list:

- **Enhanced Security:** By allocating the spread codes exclusively to authorized individuals, the disseminated information can be effectively safeguarded from unauthorized entities. Furthermore, SS methodologies can markedly diminish the likelihood of illicit receivers discerning communication patterns, thereby providing a superior level of security assurance beyond mere prevention of information leakage.
- **Resilience to Fading and Jamming:** Within the DSSS framework, fading can transpire only when the average time delay of multipath signal components is less than a single code chip duration, as multipath components extending beyond one code chip period are effectively suppressed due to the exceptional self-correlation characteristics of PRN sequences. Additionally, in FHSS systems, the agile modulation of the central frequency and the meticulous design of the frequency-hopping sequence avert the transmitted signal from frequently aligning with the fading frequency bands, thereby mitigating the probability of frequency-selective fading. Conversely, all narrow-band interference signals superimposed on the DSSS signals are transformed into wide-band interference, which can be effectively filtered out from the received valuable signal utilizing a narrow-band filter.
- **Improved Spectrum Efficiency:** Both DSSS and FHSS architectures promote spectrum utilization among multiple users within the same frequency band, constituting an effective approach to addressing interference management within satellite communication networks.

8.2.2.2 Orthogonal Time Frequency Space Modulation

1. **Definition and Introduction**

   Orthogonal time frequency space (OTFS) modulation represents a cutting-edge two-dimensional modulation framework that extends the classical time and frequency modulation methodologies, such as TDM and FDM [12]. Unlike orthogonal frequency division multiplexing (OFDM), OTFS modulation encodes information within the delay-Doppler domain, as opposed to the conventional time–frequency domain. This innovative paradigm facilitates the attainment of comprehensive time and frequency diversity, coupled with robust anti-Doppler frequency-shift capabilities.

   Moreover, OTFS modulation permits the transformation of the dynamic channel within the time–frequency domain into a static channel within the Doppler delay domain, thus mitigating the adverse effects of the time-varying characteristics inherent to the time-domain channel on overall communication efficacy. The resultant equivalent time-invariant channel located in the delay-Doppler domain exhibits beneficial attributes, such as resolvability and sparsity, which enhance effective channel estimation and signal detection, consequently improving communication dependability.

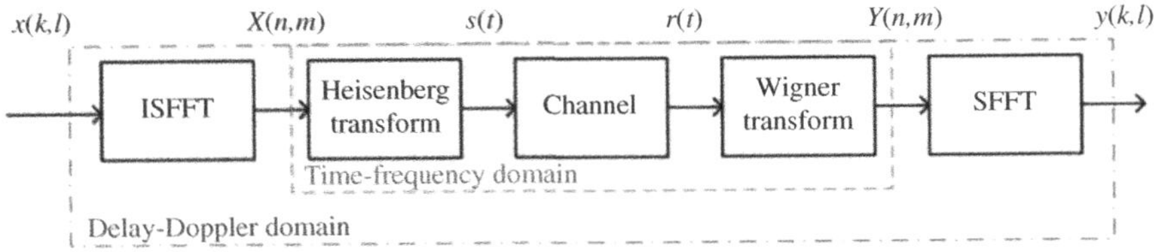

Figure 8.4 The basic architecture of OTFS.

The fundamental architecture underlying OTFS modulation is illustrated in Figure 8.4. The signal processing workflow of OTFS can be delineated as follows. At the transmitter's end, the delay-Doppler domain signal $x(k, l)$ is converted into its corresponding representation in the time–frequency domain $X(n, m)$ through the application of the inverse symplectic finite Fourier transform (ISFFT), following which the Heisenberg transform is applied to $X(n, m)$ to produce the continuous waveform $s(t)$, which is prepared for transmission. At the receiver's end, the incoming signal $r(t)$ is initially subjected to the Wigner transform in order to recover the signal within the discrete time–frequency domain $Y(n, m)$. Subsequently, $Y(n, m)$ is processed through the symplectic finite Fourier transform (SFFT) to derive the received signal $y(k, l)$ in the Doppler delay domain.

In practical applications, the suitability of OTFS for high-mobility wireless communication scenarios, such as the Internet of vehicles, high-speed rail, and UAVs, is particularly noteworthy.

2. **OTFS in Satellite Communications**

Among the rapidly evolving satellite networks, LEO satellite constellations have garnered significant international interest due to their advantages, which encompass cost-effectiveness, extensive coverage, and minimal latency. Nevertheless, the elevated operating frequencies and high orbital velocities of LEO satellites lead to pronounced Doppler effects, which adversely impact the performance of signal reception. Given the efficacy of OTFS in alleviating the detrimental impacts of the Doppler shift, the integration of OTFS within LEO satellite networks presents a viable opportunity. Compared to alternative schemes for compensating Doppler shifts, OTFS offers superior spectral efficiency, reduced implementation complexity, and a diminished peak-to-average power ratio (PAPR). The merits of utilizing OTFS in SCSs can be elucidated as follows.

1. **Enhanced Spectrum Efficiency:** OFDM facilitates efficient data transmission with augmented spectral efficiency for HTSs via multiple orthogonal subcarriers. However, the orthogonality among these subcarriers is significantly compromised in environments influenced by Doppler shifts, resulting in inter-carrier interference (ICI) and a degradation in the rate performance of OFDM. In the milieu of LEO satellite communication,

OFDM can only approximate the error performance characteristic of scenarios devoid of Doppler shift when the spacing between adjacent subcarriers exceeds ten times the maximum Doppler shift [13]. Conversely, OTFS intrinsically mitigates the limitations posed by doubly selective channels through modulation and demodulation in the Doppler delay domain, thereby producing a quasi-static and typically sparse channel conducive to satellite communications. Consequently, OTFS facilitates a more efficient selection of subcarrier spacing that is equal to or greater than the maximum Doppler shift [14]. Therefore, OTFS substantially augments the spectral efficiency of OFDM in satellite channels characterized by significant mobility.

2. **Reduced Implementation Complexity:** In the context of LEO satellite communications marked by high relative mobility, phase-locked loops (PLLs) are conventionally employed to compensate for the Doppler spread present in the received signals. However, PLLs exhibit certain shortcomings that negatively influence the overall performance of the system. On one hand, to accommodate high data rate applications utilizing multiple carriers, PLLs must track each subcarrier individually, resulting in a considerable computational burden. Moreover, the operating frequencies of the subcarriers under the influence of Doppler spread are prone to surpass the maximum permissible dynamic range of the PLLs, leading to failures in Doppler shift compensation. To mitigate these challenges, OTFS inherently obviates the Doppler shift through the implementation of delay-Doppler domain signal processing, such that only a minimal amount of residual Doppler shift requires additional compensation. Thus, both the complexity of the system and the demands placed on the dynamic range of PLLs can be significantly diminished.
3. **Lower PAPR:** It is a well-established fact that OFDM can induce elevated levels of PAPR due to the phenomenon wherein the modulated signal attains a substantial amplitude when the superposition of numerous subcarriers is in phase alignment. This phenomenon has the potential to result in the amplitude of the transmitted signal surpassing the linear operational range of the power amplifier, consequently leading to frequency distortion of the signal. Conversely, the application of OTFS modulation facilitates the dispersion of signals within the Doppler delay domain across the corresponding time–frequency domain via the ISFFT. As a result, the likelihood of all signal components constructively accumulating at a singular time instant, as observed in OFDM, is significantly diminished. Empirical studies have demonstrated that the PAPR performance exhibited by OTFS is at least comparable to that of single carrier frequency division multiple access (SC-FDMA) [15], which is the technology employed by OneWeb satellites for uplink transmission [16].

References

1 Oliver Montenbruck and Eberhard Gill. Satellite orbits: Models, methods, and applications. *Appl. Mech. Rev.*, 55(2):B27–B28, 2002.

2 Prasenjit Sengupta, Srinivas R Vadali, and Kyle T Alfriend. Satellite orbit design and maintenance for terrestrial coverage. *J. Spacecr. Rockets*, 47(1):177–187, 2010.

3 Mohamed Zayan and Fawzy Eltohamy. Orbits design for remote sensing satellite. In *2008 IEEE Aerospace Conference*, pages 1–9. IEEE, 2008.

4 Xiaoyong Du, Yang Fu, Zhengang Xue, Yuening Guo, and Jietai Mao. Simulation of the impacts of satellite orbit parameters on the number and distribution of LEO-LEO occultation events. *Chin. J. Geophys.* 50(5):1107–1116, 2007.

5 Krzysztof Sośnica. *Determination of Precise Satellite Orbits and Geodetic Parameters Using Satellite Laser Ranging*. Astronomical Institute, University of Bern, Switzerland, 2014.

6 Urs Hugentobler and Oliver Montenbruck. Satellite Orbits and Attitude. *Springer Handbook of Global Navigation Satellite Systems*. pages 59–90. Springer, 2017.

7 Thomas P Yunck. Orbit determination. *Prog. Astronaut. Aeronaut.*, 164:559–592, 1996.

8 Yue Wang, Gang Sun, and Haotong Wang. Multi-satellite task planning based on dynamic collaboration with multiple boundary constraints. In *International Conference on Signal and Information Processing, Networking and Computers*, pages 42–50. Springer, 2023.

9 Changyuan He and Yunfeng Dong. Multi-satellite observation-relay transmission-downloading coupling scheduling method. *Remote Sens.* 15(24):5639, 2023.

10 Jordi Vilà-Valls, Monica Navarro, Pau Closas, and Massimo Bertinelli. Synchronization challenges in deep space communications. *IEEE Aerosp. Electron. Syst. Mag.*, 34(1):16–27, 2019.

11 Yu Zhou, Ruhai Wang, Lei Yang, Jie Liang, Scott C Burleigh, and Kanglian Zhao. A study of transmission overhead of a hybrid bundle retransmission approach for deep-space communications. *IEEE Trans. Aerosp. Electron. Syst.*, 58(5):3824–3839, Jan. 2022.

12 Ronny Hadani, Rakib Shlomo, Tsatsanis Michail, et al. Orthogonal time frequency space modulation. In *2017 IEEE Wireless Communications and Networking Conference (WCNC)*, pages 1–6, 2017.

13 Christian Sturm, Thomas Zwick, and Werner Wiesbeck. An OFDM system concept for joint radar and communications operations. In *VTC Spring 2009–IEEE 69th Vehicular Technology Conference*, pages 1–5, 2009.

14 Zhiguo Ding, Robert Schober, Pingzhi Fan, and Vincent Poor. OTFS-NOMA: An efficient approach for exploiting heterogenous user mobility profiles. *IEEE Trans. Commun.*, 67(11):7950–7965, 2019.

15 Ronny Hadani and Anton Monk. OTFS: A new generation of modulation addressing the challenges of 5G. *arXiv preprint arXiv:1802.02623*, 2018.

16 Gilberto Berardinelli, Luis Angel Maestro Ruiz de Temino, Simone Frattasi, Muhammad Imadur Rahman, and Preben Mogensen. OFDMA vs. SC-FDMA: performance comparison in local area IMT-A scenarios. *IEEE Wirel. Commun.*, 15(5):64–72, 2008.

9

Spaceborne Payload

Spaceborne payloads denote the intricate apparatus and systems integrated within satellites that execute designated functions, serving as the essential element for the fulfillment of the satellite's operational objectives. In contrast to the satellite's fundamental subsystems, including the orbital control system, propulsion system, or power system, which predominantly uphold the satellite's functionality and stability, the payload is exclusively tasked with performing mission-specific activities. For communication satellites, the principal objective of the payload is to facilitate high-velocity data transmission between gateways (ground stations) and UE (such as mobile devices, IoT sensors, or military terminals), in addition to enabling inter-satellite communication via relay forwarding and sophisticated signal processing techniques.

The architecture of spaceborne payloads represents a profoundly intricate and interdisciplinary undertaking. It necessitates not only the assurance of effective signal processing, dynamic resource allocation, and dependable data transmission but also the resolution of the distinct challenges introduced by the inhospitable space environment. These challenges encompass severe temperature variations, exposure to radiation, impacts from micrometeoroids, and conditions of vacuum [1], all of which can materially influence the efficacy and durability of the payload's hardware. For example, components designed to withstand radiation and thermal management systems are essential for guaranteeing the operational reliability of the payload throughout its designated mission duration.

Contemporary communication satellite payloads frequently integrate cutting-edge technologies, such as software-defined radios (SDRs), PA antennas, and onboard AI-based signal processing to augment adaptability and performance. Moreover, ISLs are progressively being incorporated into payload configurations to facilitate uninterrupted data relaying among satellites in LEO, MEO, and

Safeguarding Satellite Communications: Issues, Challenges, and Solutions, First Edition.
Jianping An, et al.

GEO, thereby establishing the foundational framework of next-generation satellite constellations such as SpaceX's Starlink and OneWeb [2].

Furthermore, the design of the payload must judiciously reconcile performance, energy consumption, and mass limitations, as these variables exert a direct influence on the satellite's launch expenditures and operational efficacy. For instance, HTS [3] necessitates payloads proficient in processing and transmitting terabytes of data daily, thereby requiring the implementation of high-efficiency amplifiers, low-noise receivers, and compact, lightweight materials. Simultaneously, the payload must be architected to accommodate adaptable mission profiles, such as responding to evolving communication requirements or reconfiguring its operational parameters in real time.

The makeup of a communication satellite's payload generally comprises several pivotal components: antenna systems, signal processing units, power amplifiers, frequency conversion units, as well as data storage and processing units. Each of these components fulfills a crucial function in guaranteeing the satellite's capability to efficiently and securely receive, process, and transmit data.

- **The antenna system:** The antenna system plays a crucial role in the reception of signals emanating from ground stations, UE, or other satellites, as well as in the transmission of processed signals back to their designated recipients. Contemporary antenna designs are distinguished by their high gain and multi-beam capabilities, which facilitate extensive coverage and superior quality in signal transmission. A notable progression in antenna technology is adaptive BF, which dynamically modifies the radiation pattern of the antenna array to concentrate on signals from particular satellite links while concurrently mitigating interference from terrestrial sources or other satellites [4–6]. Through the utilization of real-time signal analysis, the antenna system is capable of optimizing its beam direction, thereby significantly enhancing signal confidentiality and diminishing the potential for passive eavesdropping and jamming. For instance, in LEO constellations, adaptive BF is imperative for addressing the swift motion of satellites in relation to the terrestrial surface, thus ensuring uninterrupted and secure communication links. Furthermore, PA antennas are increasingly being adopted due to their capability to electronically steer beams without the necessity for mechanical movement, thereby improving reliability and minimizing wear and tear.
- **The signal processing unit:** The signal processing unit constitutes the fundamental element of the payload, tasked with the demodulation, decoding, and restoration of received signals to their original formats in accordance with established communication protocols. Additionally, it is responsible for re-encoding and modulating signals for transmission, thereby ensuring precision and efficacy in communication. Sophisticated signal

processing units possess the capability to manage multiple access signals (e.g., TDMA, FDMA, CDMA) to facilitate intricate communication operations. In order to bolster security, these units frequently utilize advanced encryption algorithms, such as the advanced encryption standard (AES), which has attained widespread acceptance in real-time satellite communications owing to its efficacy and robust security features. A prominent avenue of contemporary research involves the amalgamation of super-encryption methodologies that integrate encryption with channel coding, thereby achieving both data safeguarding and bit error correction [7]. Moreover, the deployment of ML algorithms for real-time signal analysis and anomaly detection is gaining prominence, empowering the system to autonomously identify and address potential security vulnerabilities.

- **The power amplifier unit:** The power amplifier unit plays a pivotal role in augmenting signal strength to counteract spatial losses and facilitate reliable communication across extensive distances. High-efficiency amplification devices, including traveling wave tube amplifiers (TWTAs) and solid-state power amplifiers (SSPAs) [8], are frequently utilized to accomplish this objective. In order to optimize the quality of signal transmission and bolster system security, adaptive power control (APC) technology is implemented. APC systematically modifies transmission power in real time based on continuous assessment of channel conditions, such as SNR and interference levels [9]. By sustaining signal strength within an ideal range, APC alleviates the repercussions of malicious interference and diminishes the probability of unauthorized eavesdropping. Furthermore, techniques for compensating nonlinearity may be employed to reduce distortion and enhance the overall efficiency of the power amplifier [10].
- **The frequency conversion unit:** The frequency conversion unit plays a pivotal role in ensuring that signals are conveyed at the appropriate frequency by transforming the frequency band of the received signal into an appropriate transmission frequency band. This procedure is essential for the effective utilization of the frequency spectrum and adherence to international frequency allocation standards. To augment security measures, FHSS technology is frequently incorporated into the frequency conversion unit. FHSS dynamically alters the transmission frequency across a diverse array of channels, thereby complicating the efforts of adversaries to intercept or disrupt the signal [11]. When synergized with adaptive BF, FHSS markedly strengthens the robustness of satellite communications. Furthermore, SDRs are progressively being integrated into frequency conversion units [12], facilitating flexible and reconfigurable frequency management that can adapt to evolving mission requirements.

- **The data storage and processing unit:** The data storage and processing unit is responsible for the management of extensive data flows produced during the signal processing phase, facilitating the effective storage, caching, and retrieval of information. This unit holds particular significance in the context of high-throughput data management in domains such as Earth observation, remote sensing, and inter-satellite communication. To uphold data integrity and security, cryptographic hash algorithms (e.g., SHA-256 [13]) are employed to generate hash values that verify data authenticity, thus enabling the immediate identification of any tampering. Furthermore, data integrity verification mechanisms such as cyclic redundancy checks (CRCs) [14] and redundant storage methodologies, including redundant arrays of inexpensive disks (RAID) technology [15], are utilized to mitigate data corruption risks and to ensure the continuous availability of data in the occurrence of hardware malfunctions. The incorporation of onboard AI-driven data processing is also emerging as a pivotal trend, facilitating instantaneous data analysis and autonomous decision-making independent of terrestrial systems. For instance, AI algorithms may be deployed to prioritize data transmission based on criticality or to identify and classify anomalies within the data stream.

Through the synergistic functioning of these integral components, the payloads of communication satellites are capable of executing functions such as signal reception, processing, and transmission with both efficiency and stability, thereby facilitating the seamless operation of the SCS. As the imperative for global connectivity and secure communication escalates, innovations in payload technology will assume a critical role in determining the future trajectory of satellite-based communication infrastructures.

Spaceborne payloads can be systematically classified into OBP payloads and transparent forwarding payloads (commonly known as regenerative payloads) according to their functional architectures. These two categories of payloads align with disparate satellite communication network architectures: the space-based network architecture and the ground-based network architecture [16]. Each architecture manifests distinct characteristics and is employed in contexts with particular operational exigencies, including latency sensitivity, throughput requirements, and security factors. The ensuing sections of this chapter will deliver a thorough examination of these two payload categories, emphasizing three principal dimensions: operating principles, intrinsic characteristics and design considerations, and security technologies. It is also significant to note that certain advanced satellite systems amalgamate both OBP and transparent forwarding functionalities within a singular payload, facilitating dynamic mode transitions to accommodate a variety of communication tasks and operational contexts.

9.1 On-board Processing

9.1.1 Operating Principles

OBP payloads are predominantly utilized within space-based network architectures [17], which exploit the inherent capabilities of satellites to conduct signal processing and routing operations in the orbital environment. As illustrated in Figure 9.1, the UE establishes a connection with the satellite through a designated service link. Upon the reception of the signal, the OBP payload executes essential functions, including demodulation, decoding, encryption, and re-encoding. The processed signal is subsequently relayed via ISLs through one or more hops to a satellite situated in closer proximity to the gateway. This satellite remodulates the signal and disseminates it to the gateway via a feeder link. Ultimately, the gateway amplifies, re-encodes, or forwards the signal to the core network, thereby completing the communication continuum.

This architectural framework is distinctly adept for LEO satellite constellations, which are defined by their minimal latency, substantial data throughput, and swift orbital dynamics. LEO satellites, functioning at altitudes typically varying from

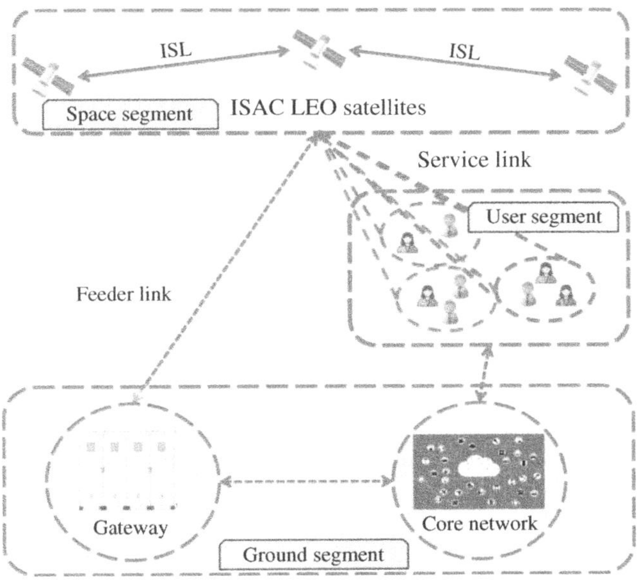

Figure 9.1 The space-based network architecture.

500 to 2,000 km [17], furnish a robust platform for the deployment of OBP payloads. These payloads are integrated with sophisticated DSP units that facilitate the conversion of frequency-modulated signals into digital formats, the restoration of the original data, and the implementation of encryption algorithms to safeguard data confidentiality and integrity. Prior to transmission, the payloads further encode and modulate the data to enhance its optimization for diverse channel conditions, thereby augmenting its resilience against interference and noise.

The brief orbital period and elevated mobility of LEO satellites empower them to provide low-latency, real-time data processing, rendering them particularly suitable for applications including real-time Earth observation, military communications, and emergency response systems. The interconnected configuration of LEO constellations, enabled by laser-based ISLs or microwave radio connections, facilitates rapid data exchange among satellites. This functionality guarantees that data can be proficiently routed through the network, even in scenarios devoid of direct visibility to ground stations. For instance, laser communication systems present high bandwidth and a low probability of interception, thus positioning them as particularly advantageous for secure military and governmental communications [18].

In addition to LEO satellites, GEO satellites can assume a supplementary function within space-based network architectures. Despite the higher latency constraints imposed by their altitude of 36,000 km [17], GEO satellites offer global coverage and stable connectivity [19], rendering them indispensable for applications such as key management, global network monitoring, and disaster recovery communications. GEO satellites can act as relay nodes for LEO constellations, managing long-distance communication tasks and ensuring network resilience. In circumstances where LEO satellites encounter failures or disruptions in ISLs, GEO satellites can function as backup nodes, preserving communication continuity and averting service interruptions.

The amalgamation of hybrid satellite networks—incorporating both LEO and GEO satellites—signifies a noteworthy progression in space-based communication systems [20]. For instance, GEO satellites can furnish persistent surveillance and command-and-control functionalities, while LEO satellites facilitate high-speed, low-latency data transmission. This synergistic interaction enhances the comprehensive performance and reliability of the network, particularly within mission-critical applications such as defense operations, maritime surveillance, and global IoT connectivity.

9.1.2 Characteristics and Design Considerations

The distinctiveness of OBP payloads resides in their capacity to transfer the preponderance of data processing responsibilities from terrestrial stations to orbital platforms, thereby metamorphosing satellites into sophisticated nodes equipped with superior signal processing and data management functionalities. In contrast to conventional transparent forwarding payloads, which are predominantly reliant on ground-based infrastructure for signal regeneration, OBP payloads amalgamate various processing phases directly within the satellite, encompassing signal demodulation, decoding, encryption, and routing. This integration empowers the satellite to autonomously execute the entire procedure, from signal reception to processing and transmission, thereby markedly diminishing reliance on ground stations and abbreviating the signal processing trajectory [16].

The efficacy of OBP payloads is primarily ascribed to their robust real-time data processing proficiencies. Furnished with high-performance DSPs [21] and FPGAs [22], these payloads are capable of intelligently modulating routing in accordance with real-time data influx, thereby ensuring that high-priority information is conveyed to its intended destination with minimal latency. For example, in LEO constellations, satellites possess the capability to dynamically redirect data through ISLs to circumvent congested or compromised pathways, thus enhancing network performance. This adaptive routing mechanism not only augments the overall efficiency of the satellite communication network but also guarantees low latency and high reliability, rendering it particularly advantageous for applications necessitating rapid responsiveness, such as military communications, disaster recovery, and autonomous vehicle navigation.

To further fortify the fault tolerance and resilience of OBP payloads, a redundant design is implemented at both hardware and software tiers. Essential processing components, such as signal demodulators and encryption modules, frequently employ dual or triple redundancy, thereby assuring uninterrupted functionality even in the circumstance of a component malfunction. For instance, if a processing unit experiences failure due to radiation-induced anomalies, a secondary unit can promptly assume control, preserving seamless communication. Moreover, error detection and correction algorithms, such as FEC [23] and CRC, are embedded within the payload to identify and rectify data loss or corruption in real time. These mechanisms are particularly vital in the inhospitable space environment, where cosmic radiation and thermal cycling can impair system performance.

The adaptability of OBP payloads represents an additional salient advantage. These payloads are capable of modifying their processing methodologies in accordance with specific mission imperatives, such as prioritizing low-latency communication for time-sensitive operations or optimizing bandwidth utilization for high-throughput endeavors. For instance, in multi-beam satellite architectures, the payload can dynamically allocate resources to various beams predicated on traffic demand [24], thereby ensuring the efficient allocation of available bandwidth. This highly adaptable framework empowers satellites to sustain efficient and reliable operations even amid intricate and fluctuating space environments.

In conclusion, OBP payloads facilitate contemporary SCSs in significantly alleviating their dependence on terrestrial stations while augmenting data processing capacity, flexibility, redundancy, and security. These attributes render OBP payloads fundamental components of next-generation satellite networks, ensuring efficient and stable operation across diverse and challenging space environments. The subsequent sections will delve deeper into the technical principles, design considerations, and security technologies associated with these payloads, thereby providing a thorough comprehension of their role in the advancement of satellite communications.

9.1.3 Security Technologies

9.1.3.1 Satellite-to-ground Secure Transmission Based on IPSec

Space-ground collaboration constitutes a fundamental characteristic of the space-ground network architecture, which synthesizes a cohesive network framework, standardized systems, and sophisticated technologies such as wireless access, transmission protocols, and network management infrastructures. This architecture implements a collaborative resource allocation and service management framework between space and ground entities, thereby facilitating seamless coordination between space-based and terrestrial systems [25]. By capitalizing on this strategic framework, the network attains comprehensive coverage for users, guaranteeing uninterrupted communication services even in remote or geographically challenging environments. This capability is particularly pivotal for applications including global IoT connectivity, emergency response frameworks, and military operations, wherein continuous and reliable communication is of paramount importance.

At the core of space-ground collaboration is space-ground backhaul technology, which enables the transmission of data from OBP payloads to ground gateways through feeder links, ultimately relaying the data to the ground-core network. This technology is characterized by several distinctive features that set it apart from conventional terrestrial backhaul systems. Firstly, the communication link connecting OBP payloads and ground gateways is typically a single-point link, necessitating precise alignment and synchronization to sustain connectivity. Secondly, the space-ground backhaul must accommodate substantial volumes of communication data, ensuring that the bandwidth requirements for the interface between the payload and the core network are adequately fulfilled. This aspect is particularly critical for high-throughput applications, such as Earth observation [26], broadband Internet services [27], and real-time video streaming [28].

Furthermore, the extensive communication distances inherent in space-ground backhaul, often extending across thousands of kilometers, demand robust signal amplification and error correction mechanisms to preserve data integrity. The system must also be equipped to manage the high-speed mobility of satellites, particularly in LEO constellations, where satellites exhibit rapid movement relative to the ground. To mitigate this challenge, space-ground backhaul links are engineered to support real-time handover between different gateways, thereby ensuring uninterrupted communication as satellites navigate their orbits. This handover mechanism is predicated on advanced BF techniques [29] and predictive tracking algorithms [30] to forecast satellite trajectories and optimize link performance.

In the realm of secure data transmission, the IPSec protocol functions as a principal security mechanism for space-ground backhaul, safeguarding the confidentiality, integrity, and authenticity of data exchanged between the satellite payload and the ground core network. IPSec facilitates end-to-end encryption (E2EE) and authentication, thereby protecting sensitive information from interception or tampering during the transmission process. Depending on specific design imperatives, two primary IPSec transmission schemes are employed for space-ground backhaul: Payload IPSec and gateway station IPSec.

- **Payload IPSec:** In the Payload IPSec framework, the satellite payload directly initiates an IPSec tunnel with the terrestrial core network, thereby guaranteeing E2EE and secure data transmission. This framework necessitates that the satellite payload dynamically acquire a public IP address via the gateway, which is subsequently employed to establish the IPSec tunnel with the secure gateway of the core network. This methodology affords a substantial degree of security, as the data remains encrypted throughout the entire transmission pathway, extending from the satellite to the core network. Nevertheless, this framework encounters considerable challenges in circumstances where the satellite is subject to continuous high-velocity movement, as observed in LEO constellations. Owing to the swift displacement of the satellite, the gateway connection necessitates frequent transitions, compelling the payload to repeatedly acquire a public IP address and re-establish the IPSec tunnel. This procedure, while upholding security, introduces latency and complexity in the management of the tunnel. As illustrated in Figure 9.2, the business data is encapsulated with IPSec during transmission, thereby ensuring its encryption and secure conveyance to the terrestrial core network. The dynamic formation of the IPSec tunnel constitutes a pivotal characteristic of this framework, enabling it to adapt to the intricate space-ground transmission milieu. However, the recurrent re-establishment and switching of tunnels may exacerbate implementation complexity and operational overhead, rendering this framework more conducive to scenarios characterized by infrequent handovers or where point-to-point security is of paramount importance.

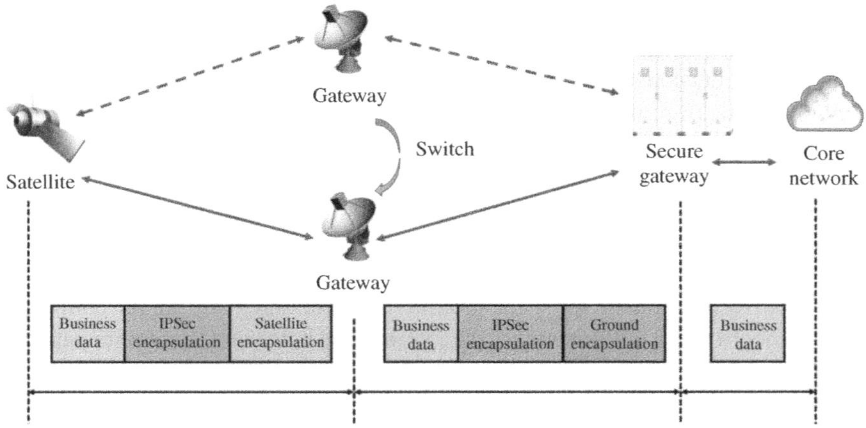

Figure 9.2 A secure communication scheme for OBP payload based on Payload IPSec.

- **Gateway IPSec:** In contrast to the Payload IPSec methodology, the gateway IPSec paradigm centralizes the initiation of the IPSec tunnel at the gateway tier. In this framework, the gateway directly formulates an IPSec tunnel with the terrestrial core network, while the satellite payload interfaces with the gateway via a preassigned internal IP address. This configuration streamlines the communication mechanism for the satellite payload, as it is solely required to relay data to the gateway through the feeder link, thereby entrusting the responsibilities of encryption and secure transmission to the gateway. Subsequently, the gateway transmits the data to the core network through the preestablished IPSec tunnel. This architecture effectively insulates the satellite payload from the intricacies associated with tunnel management, thereby enhancing the efficiency of the switching process. During instances of gateway handovers, the satellite payload is only necessitated to re-establish the link-layer connection to the newly designated gateway, without the requirement to re-initiate the IPSec tunnel. This markedly diminishes switching latency and augments the overall performance of the system. As depicted in Figure 9.3, the gateway is tasked with the reception and encryption of the satellite payload's business data, subsequently facilitating secure communication with the core network via the IPSec tunnel. Furthermore, the augmented air interface security mechanisms of the feeder link, such as FH and adaptive BF, afford substantial protection against vulnerabilities such as air interface interference, eavesdropping, forgery, and hijacking. These mechanisms further fortify the security and reliability of the data transmission, rendering the Gateway IPSec scheme particularly advantageous for scenarios that necessitate frequent handovers and minimal switching latency.

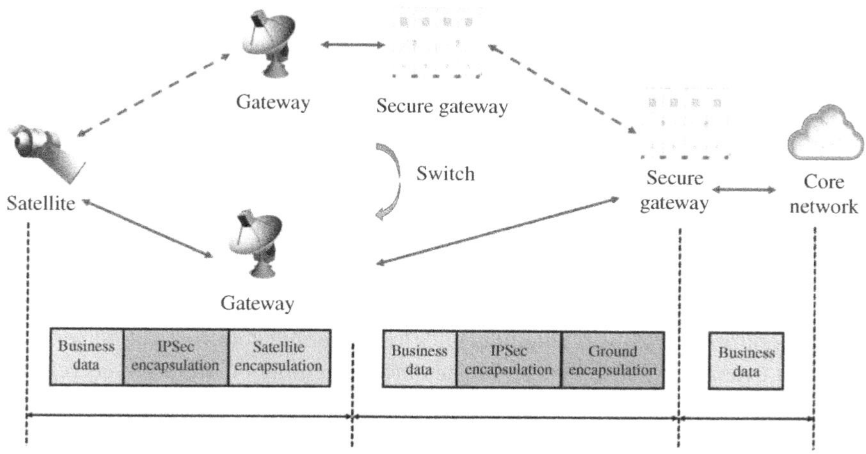

Figure 9.3 A secure communication scheme for OBP payload based on Gateway IPSec.

These two architectures demonstrate divergent characteristics and are tailored for distinct operational contexts. The Payload IPSec architecture provides versatility and direct end-to-end security, rendering it particularly advantageous for applications where point-to-point communication is paramount and handovers are infrequent. Nevertheless, its dependence on the dynamic re-establishment of tunnels may introduce heightened complexity and latency in environments characterized by high mobility, such as LEO constellations. Conversely, the Gateway IPSec architecture streamlines the switching process via centralized tunnel management, thereby mitigating latency and operational overhead. This renders it especially suitable for scenarios that entail frequent handovers and high-velocity satellite mobility, such as global broadband networks and real-time Earth observation systems. Additionally, the Gateway IPSec architecture's reliance on advanced air interface security mechanisms enhances the resilience of the feeder link against potential threats, thereby ensuring robust and secure data transmission.

In conclusion, the decision-making process regarding the selection between Payload IPSec and Gateway IPSec should be informed by the specific requirements inherent to the space-ground backhaul scenario. Payload IPSec is distinguished by its ability to provide flexible, end-to-end secure communication, whereas Gateway IPSec delivers efficient switching performance and enhanced link security through centralized management. By meticulously assessing factors such as handover frequency, latency tolerance, and security prerequisites, system designers can identify the most suitable architecture to optimize the performance and reliability of their satellite communication networks.

9.1.3.2 ISL Secure Transmission Based on Laser Communication

ISLs constitute essential conduits for communication that interlink the payloads of satellites, the core network of the satellite system, and satellites positioned in medium-to-high orbits, thereby establishing a resilient extraterrestrial communication infrastructure. These ISLs facilitate high-velocity data transmission and instantaneous data interchange among multiple satellites, effectively converting them into nodes of exchange within a decentralized network framework [31]. By capitalizing on ISLs, satellite systems can realize dynamic communication capabilities and collaborative operations in the space environment, thus optimizing data throughput and augmenting network dependability. This cohesive architecture not only empowers satellites to deliver regional communication services but also extends global coverage, particularly in remote or underserved regions where conventional terrestrial infrastructure is either nonexistent or impractical to implement.

The incorporation of laser communication technology into ISLs signifies a substantial enhancement over conventional microwave communication modalities [32]. Laser communication presents several distinct benefits, such as extraordinarily high data transfer rates, resilience to limitations of the wireless spectrum, and improved security features. These characteristics render laser communication an optimal solution for contemporary satellite networks, especially in applications that necessitate high bandwidth and minimal latency, including expansive IoT implementations, high-definition video dissemination, and real-time data analytics. Furthermore, laser communication systems are characterized by robust anti-jamming capacities, reduced power consumption, and elevated confidentiality, positioning them as a preferred option for military communications and secure data exchanges. These attributes establish laser communication as a fundamental element of forthcoming satellite communication frameworks [33].

Notwithstanding its benefits, laser communication encounters several technical obstacles, particularly in contexts involving satellites situated in disparate orbits. The elevated relative velocities of satellites, coupled with the necessity for precise beam alignment, inject considerable complexity into the deployment and upkeep of laser communication systems [34]. For instance, within LEO constellations, satellites exhibit rapid movement relative to GEO satellites, necessitating continuous realignment efforts to sustain beam alignment. To surmount these challenges, advanced optical alignment systems and automated adjustment technologies are imperative. Such systems depend on high-precision actuators, real-time tracking algorithms, and adaptive optics to guarantee stable and precise beam alignment. Additionally, intelligent network management systems are pivotal in dynamically optimizing link performance, facilitating seamless transitions, and minimizing latency.

A principal challenge in laser communication involves attaining efficient and stable connectivity among satellites situated in heterogeneous orbits, such as LEO and GEO [35]. The extended communication distances and diverse orbital dynamics demand the implementation of intricate alignment mechanisms and dynamic adjustment protocols. For instance, predictive tracking algorithms can forecast satellite trajectories and preemptively modify beam orientations [30], while error correction methodologies can mitigate alignment inaccuracies induced by environmental variables such as thermal variations and mechanical oscillations. Moreover, the integration of ML and AI into network management systems facilitates autonomous optimization of link performance, thereby ensuring reliable communication even in highly variable environments.

In conclusion, ISLs, especially those that incorporate laser communication technologies, are fundamentally transforming the functionalities of satellite networks. By facilitating high-velocity data transmission, universal coverage, and adaptive collaboration, ISLs are laying the groundwork for advanced applications, including global IoT connectivity, instantaneous terrestrial observation, and secure military communications. Nevertheless, the effective deployment of laser-based ISLs necessitates the resolution of substantial technical obstacles, such as accurate beam alignment, real-time adjustment systems, and sophisticated network management. Tackling these obstacles will be imperative for realizing the complete potential of laser communication within satellite networks and for assuring their dependability and efficacy across a variety of operational contexts.

9.1.3.3 Blockchain Technology

The incorporation of blockchain technology into OBP payloads signifies a revolutionary methodology for augmenting the security of satellite communications. By harnessing the fundamental attributes of blockchain, such as immutability, decentralization, and transparency, satellite systems can effectively mitigate significant security threats, including data tampering, unauthorized access, and identity spoofing and simultaneously enhancing the overall resilience of the network [36]. These attributes furnish substantial assurances regarding the integrity, authenticity, and confidentiality of data processed and conveyed by satellite payloads, rendering blockchain an essential instrument for safeguarding space-based communication systems.

The onboard payload is tasked with the processing and management of essential data, encompassing Earth observation imagery, environmental monitoring data, and scientific measurements. Throughout its operation, the payload generates data summaries or metadata that are subsequently inscribed onto the blockchain. The immutable characteristic of blockchain guarantees that any unauthorized alterations to the data are readily identifiable, thus preserving the authenticity and integrity of the information [37]. This functionality is particularly critical in mission-essential applications, such as military reconnaissance and disaster monitoring, where the accuracy and reliability of data are of utmost importance. By leveraging blockchain technology, satellite systems can ensure that the data transmitted to ground stations or other satellites remains unaltered or fabricated, thereby significantly enhancing the credibility and trustworthiness of the information.

Conventional communication between satellites and gateways or other satellites frequently depends on centralized management systems, which are susceptible to cyberattacks, such as distributed DDoS attacks [38] or man-in-the-middle (MITM) attacks [39]. Blockchain technology mitigates these vulnerabilities by facilitating decentralized trust mechanisms. Through blockchain, satellites can initiate secure peer-to-peer communication independent of a central authority. Each participant within the network is authenticated through cryptographic protocols, which thwart identity spoofing and guarantee that only authorized entities are permitted to engage in data exchange [40, 41]. This decentralized paradigm not only fortifies the security of data transmission but also enhances the system's resilience to interference and eliminates single points of failure.

In addition to fortifying data integrity and communication, blockchain augments the transparency and auditability of satellite operations. All data exchanges, operational logs, and communication protocols may be inscribed onto the blockchain, thereby establishing a tamper-proof ledger accessible to all authorized participants [37]. This transparency guarantees that every action performed during data processing and transmission is traceable and verifiable, thereby diminishing the risk of data manipulation or malicious activities. For instance, within multi-stakeholder satellite missions, such as international Earth observation programs, blockchain can furnish a transparent platform for the sharing of data and resources, thereby nurturing trust and collaboration among participants.

Another notable benefit of blockchain technology lies in its facilitation of smart contracts, which serve to automate security protocols and operational workflows. Smart contracts represent self-executing programs that operate on the blockchain, enforcing established rules devoid of intermediary involvement. In the realm of OBP payloads, smart contracts can be implemented for the management of encryption keys, control of data access, and verification of compliance [42]. For example, when a satellite conveys data to a ground station, a smart contract can autonomously authenticate the credentials of the recipient, impose encryption standards, and document the transaction within the blockchain. This automation not only enhances operational efficiency but also mitigates security vulnerabilities associated with manual interventions, such as human error or insider threats.

Moreover, the distributed storage functionalities of blockchain confer an additional layer of security upon satellite systems. By dispersing critical data across numerous nodes within the network, the blockchain guarantees data redundancy and accessibility, even amid hardware malfunctions or cyberattacks [43]. This distributed framework is particularly advantageous for prolonged missions or remote operations, where access to centralized infrastructure may be constrained.

The incorporation of blockchain technology into OBP payloads presents a holistic approach to augmenting the security and reliability of satellite communications. By offering assurances of data integrity, decentralized trust frameworks, automated smart contracts, and transparent platforms for data exchange, the blockchain empowers satellite systems to operate with both efficiency and security in the face of diverse threats and challenges. This guarantees the confidentiality, integrity, and availability of data throughout complex satellite missions, thereby significantly enhancing the overall security and resilience of space-based communication networks.

9.2 Transparent Forwarding

9.2.1 Operating Principles

Transparent forwarding payloads are predominantly utilized in terrestrial network frameworks, wherein the transponders of the satellite operate as transparent relays [17]. In contrast to OBP payloads that execute intricate operations such as signal demodulation, decoding, re-encoding, and routing, the functionality of transparent forwarding payloads remains comparatively uncomplicated. These payloads principally concentrate on signal amplification and frequency conversion, directly transmitting the received signals to the designated station without modifying the data content. This methodology depends on the satellite's transponder, which comprises essential components including power amplifiers and frequency converters, to facilitate efficient and prompt signal transmission. The primary benefit of transparent forwarding payloads resides in their minimal latency and streamlined architecture, rendering them particularly advantageous for applications necessitating elevated real-time performance and rapid processing capabilities, such as broadcast services, live video streaming, and expansive communication networks.

As illustrated in Figure 9.4, the unprocessed signal acquired by the satellite undergoes amplification and frequency conversion prior to being transmitted unaltered to the feeder link, which subsequently directs the signal to the gateway. During this procedure, no decoding, reorganization, or data processing is performed, thereby considerably diminishing the complexity of the system. This optimized methodology precludes the necessity for intricate OBP operations, permitting the transparent forwarding payload to concentrate on ensuring stable signal transmission while providing a straightforward yet effective communication pathway. For instance, in direct-to-home (DTH) broadcasting [44], transparent

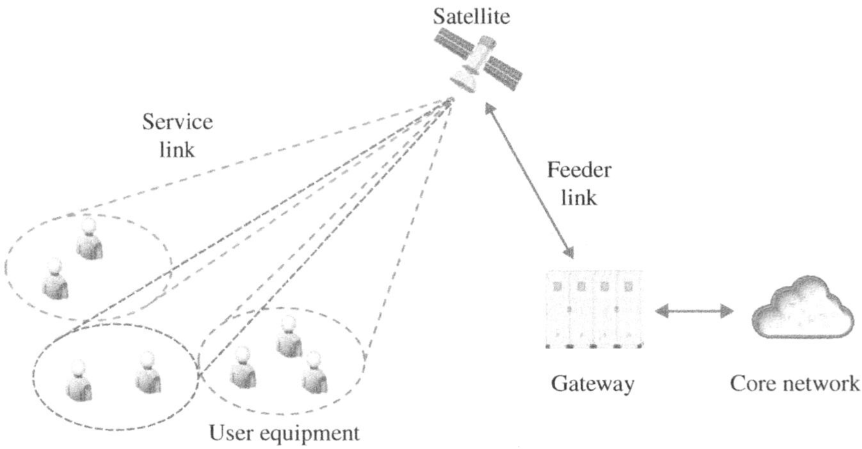

Figure 9.4 The ground-based network architecture.

payloads facilitate the seamless distribution of high-quality audio and video signals to end-users without introducing supplementary processing latencies.

The architecture of transparent forwarding payloads deliberately avoids intricate computations or alterations of data, culminating in high bandwidth efficacy and the capacity to manage substantial data volumes. This renders transparent forwarding payloads markedly more direct and efficient with respect to processing capabilities in contrast to OBP payloads [45]. By streamlining the design of the satellite communication link, transparent forwarding payloads accomplish effective data transmission, thereby minimizing the likelihood of delays or data loss during transit. For example, in maritime and aeronautical communications, where the immediacy of data delivery is crucial, transparent forwarding payloads guarantee that information is communicated swiftly and dependably.

Nonetheless, the inherent simplicity of transparent forwarding payloads also engenders specific constraints. These payloads are devoid of the capability to execute complex data processing or protocol conversions and are primarily utilized as signal-relay apparatuses. Consequently, although transparent forwarding payloads excel in communication efficiency and real-time responsiveness, they are less appropriate for contexts necessitating advanced data processing or particular communication protocols. For example, in secure military communications or scientific missions involving data analysis, the lack of OBP capabilities may constitute a significant limitation. Nevertheless, the low-latency and high-throughput characteristics of transparent forwarding payloads render them essential for applications where rapidity and simplicity are of utmost importance.

9.2.2 Characteristics and Design Considerations

The attributes of transparent forwarding payloads are fundamentally characterized by their streamlined signal transmission and elevated operational efficacy. The primary objective of these payloads is to guarantee that signals obtained from UE are expediently and reliably relayed to the gateway, without engaging in intricate processing. As a result, the architecture of transparent forwarding payloads is concentrated on enhancing the signal transmission pathway, minimizing latency, and maximizing transmission efficacy. These characteristics render transparent forwarding payloads especially advantageous for applications necessitating real-time data dissemination, such as broadcasting services, live video streaming, and extensive communication networks.

The streamlined architecture of transparent forwarding payloads culminates in diminished hardware complexity when contrasted with OBP payloads. Given that the signals do not necessitate decoding, decryption, or re-encoding, the satellite hardware is relieved of substantial computational burdens. This decrement in processing requirements yields lower power consumption and diminished system complexity, thereby rendering transparent forwarding payloads more economical viable and simpler to implement. From a hardware standpoint, the design of transparent forwarding payloads is predominantly centered around optimizing the RF module and ensuring the reliability of the signal-forwarding pathway. Essential components, including high-efficiency power amplifiers and low-noise frequency converters, are meticulously selected to uphold signal integrity and curtail losses throughout transmission. Within this design paradigm, the satellite operates chiefly as a signal relay node, ensuring that signals are transmitted with negligible delay and devoid of superfluous processing overhead.

A pivotal design consideration for transparent forwarding payloads is the maximization of bandwidth utilization, particularly in LEO and MEO satellite systems, where communication links are frequently hampered by restricted bandwidth resources. To surmount this challenge, transparent forwarding payloads can integrate multiple access technologies, such as TDMA, FDMA, or CDMA. These technologies facilitate the efficient management of multiple signal streams within the same frequency band or temporal window, enabling the payload to accommodate concurrent transmissions for numerous users. For instance, in a TDMA-based framework, time slots are allocated to various users, guaranteeing that each user has exclusive access to the communication channel without interference. Similarly, FDMA partitions the available frequency spectrum into smaller sub-bands, while CDMA employs unique codes to differentiate between users. By leveraging these technologies, transparent forwarding payloads can significantly enhance bandwidth utilization, avert resource depletion, and augment the overall transmission capacity of the system. This capability is particularly advantageous in contexts characterized by elevated data volumes or high user densities, such as urban broadband networks or maritime communication systems.

Despite the advantages associated with their design, the streamlined architecture of transparent forwarding payloads concurrently presents considerable security vulnerabilities. Given that these payloads lack OBP capabilities and data examination, they remain susceptible to exploitation by malevolent entities. For example, adversaries may take advantage of the absence of signal authentication to inject unauthorized information into the communication channel, which is subsequently relayed by the satellite undetected. This particular vulnerability raises significant concerns within military or governmental communication frameworks, wherein the preservation of data integrity and confidentiality is of utmost importance. Furthermore, transparent forwarding systems typically necessitate an extensive network of ground segments to achieve comprehensive global coverage, thereby creating a broad attack surface that could be targeted by potential adversaries [16]. For instance, an assailant might focus on ground stations to intercept or manipulate data, thereby jeopardizing the integrity of the entire communication network.

In conclusion, the attributes of transparent forwarding payloads are primarily concentrated on the simplification of signal transmission and the enhancement of communication efficacy. The design principles emphasize the optimization of the signal link, augmentation of transmission rates, minimization of latency, and maximization of bandwidth utilization through sophisticated multiple access mechanisms. These enhancements empower transparent forwarding payloads to deliver stable and rapid communication services, rendering them particularly suitable for real-time applications and extensive user access. Nonetheless, the simplified architecture of such payloads concurrently gives rise to significant security threats, necessitating the incorporation of robust encryption, authentication, and monitoring systems to safeguard the integrity and confidentiality of the transmitted information. By effectively addressing these challenges, transparent forwarding payloads can maintain their essential function within contemporary satellite communication infrastructures.

9.2.3 Security Technologies

9.2.3.1 QKD Technology

Through the implementation of QKD technology, diverse nodes within the terrestrial network architecture are able to securely acquire shared keys, thereby significantly mitigating the potential threats associated with key leakage and eavesdropping, while simultaneously safeguarding the confidentiality and integrity of data throughout the transmission process [46]. In this context, the terrestrial QKD network furnishes each node with a secure and dependable quantum key, thereby ensuring robust security throughout the data transmission phase.

As illustrated in Figure 9.5, quantum encryption gateways and specialized satellite terminals can be strategically deployed between UE and satellites to facilitate secure communication. The quantum encryption gateway retrieves quantum keys along with their corresponding identifiers from the QKD network via the quantum key infusion platform, which guarantees the dynamic updating and secure management of keys. Furthermore, quantum security functionalities are embedded within the quantum base station gateway to facilitate interaction with the QKD network and to acquire the requisite quantum keys and identifiers.

The quantum encryption gateway, functioning as a sophisticated high-performance server apparatus, possesses not only the capability to securely store and infuse quantum keys but also exhibits the flexibility to augment its storage capacity in accordance with specific business requirements and to dynamically administer the keys in relation to their usage frequency and established update protocols. Quantum keys may be employed directly as session keys or as encryption keys for session keys. Through the application of quantum encryption, session keys are transmitted with a high degree of security, thus ensuring that both communicating parties can effectively synchronize their session keys. The volume of quantum keys infused during each session is contingent upon the precise business necessities [47].

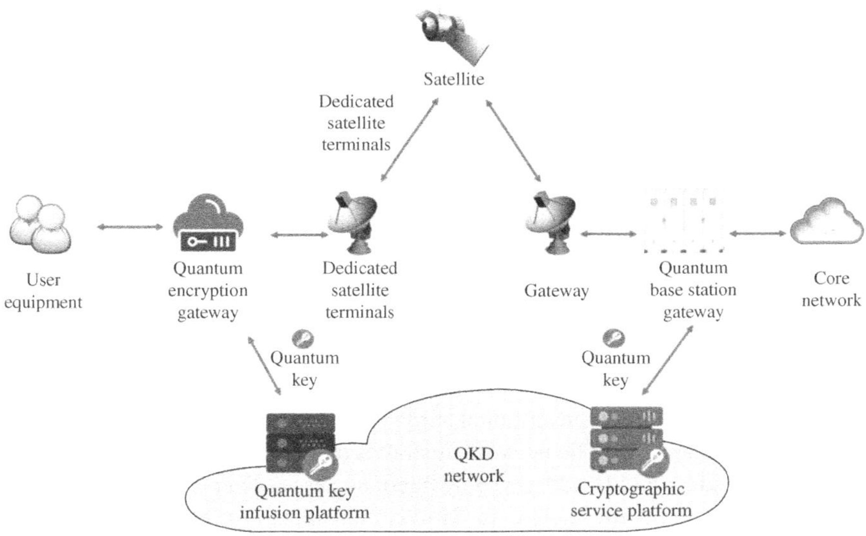

Figure 9.5 A secure communication scheme for transparent forwarding payload based on QKD.

Subsequent to the infusion of quantum keys into the quantum encryption gateway, a distinctive key identifier is generated for each individual quantum key. When the UE initiates a communication protocol, the data is initially aggregated and routed to the quantum encryption gateway. The quantum encryption gateway subsequently encrypts the data in accordance with the predefined quantum key usage and update protocols, utilizing the infused quantum key, and transmits the encrypted data along with the key identifiers through dedicated satellite terminals to the satellite. The satellite employs its transparent forwarding payload to seamlessly forward the encrypted data and key identifiers to the gateway. Upon reaching the gateway, the data and key identifiers are relayed through a dedicated line to the quantum base station gateway. Following the reception of the key identifier, the quantum base station gateway either retrieves the key locally or solicits it from the cryptographic service platform. Once the key has been acquired, the encrypted data is decrypted and forwarded to the core network, thereby ensuring comprehensive end-to-end security encryption of user data. This methodology fully exploits the attributes of transparent forwarding payloads, thus guaranteeing that every segment of data transmission from the user device to the ultimate service system remains encrypted and protected, thereby upholding the overarching security of the network and the integrity of the data [48].

9.2.3.2 Identity Authentication Technology

In order to proficiently tackle the challenge of device identity authentication within terrestrial network architecture, the implementation of advanced post-quantum cryptography (PQC) technology is advocated for authentication. Specifically, devices situated on the user side, the satellite side, and the ground network side are capable of utilizing standardized PQC digital signature algorithms such as Dilithium [49], FALCON [50], and SPHINCS+ [51] to execute the signing and verification of device identities prior to the establishment of a connection, thereby ensuring that the identities of both communicating entities are duly authenticated.

In the context of executing signature verification among devices, signature algorithms are conventionally employed to authenticate device-specific identity information (such as MAC addresses, device IDs, and device models) to ascertain the authenticity and validity of the identity data. Nevertheless, device identity encompasses not merely conventional hardware characteristics but also extends to network characteristics. The network attributes of a device (including fully qualified domain name and IP address) may fluctuate across diverse network environments, rendering reliance solely on hardware attributes for identity verification inadequate. To further augment the precision and integrity of the authentication process, it is imperative that network attributes are integrated into the authentication framework. This integration can effectively mitigate authentication failures or potential security threats arising from variations in the device's network environment.

Moreover, in order to guarantee the distinctiveness of each device signature, avert replay attacks, and ensure the timeliness of device identity authentication, timestamp information can be incorporated during the device identity signing process [52]. Timestamps not only bolster the temporal relevance of the signature, thus preventing the reutilization of historical identity information, but also ensure that each authentication operation remains independent and unique in the context of cross-time-zone or cross-domain network architectures. Consequently, by amalgamating the device's hardware attributes, network attributes, and timestamp information, the distinctiveness and immediacy of each device identity verification are assured, thereby significantly enhancing the reliability and security of the authentication procedure. This authentication framework provides robust security for terrestrial network architecture, effectively addressing the threat of transponder theft and offering a more substantial foundation of trust for communication within intricate network environments.

References

1 Herman Mark, Ralph D Soomers, and Michael J Mirtich. Effect on surface thermal properties of calibrated exposure to micrometeoroid environment. *AIAA J.*, 4(10):1811–1817, 1966.

2 Nils Pachler, Inigo del Portillo, Edward F Crawley, and Bruce G Cameron. An updated comparison of four low earth orbit satellite constellation systems to provide global broadband. In *2021 IEEE International Conference on Communications Workshops (ICC Workshops)*, pages 1–7, 2021.

3 Hector Fenech, Sonya Amos, Alessia Tomatis, and Viphakone Soumpholphakdy. High throughput satellite systems: An analytical approach. *IEEE Trans. Aerosp. Electron. Syst.*, 51(1):192–202, 2015.

4 Ye-Eun Chi, Jinki Park, and Seong-Ook Park. Hybrid multibeamforming receiver with high-precision beam steering for low earth orbit satellite communication. *IEEE Trans. Antennas Propag.*, 71(7):5695–5707, 2023.

5 Hongpeng Zhu, Cong Li, Jian Cheng, Jing Hu, and Guangxia Li. Transmit power optimization and precoding design in multiuser satellite MIMO downlink with SINR constraints. *IEEE Internet Things J.*, 10(12):10547–10558, 2023.

6 Sareh Majidi Ivari, Nazli Ahmad Khan Beigi, Mohammad Reza Soleymani, and Yousef Shayan. On optimal power allocation in multibeam multicast NOMA for satellite communication systems. *IEEE Trans. Aerosp. Electron. Syst.*, 59(5):7233–7251, 2023.

7 Suhyeon Jeon, Jeongho Kwak, and Jihwan P Choi. Cross-layer encryption of CFB-AES-TURBO for advanced satellite data transmission security. *IEEE Trans. Aerosp. Electron. Syst.*, 58(3):2192–2205, 2021.

8 Whitney Q Lohmeyer, Raichelle J Aniceto, and Kerri L Cahoy. Communication satellite power amplifiers: Current and future SSPA and TWTA technologies. *Int. J. Satell. Commun. Netw.*, 34(2):95–113, 2016.

9 Anestis Tsakmalis, Symeon Chatzinotas, and Björn Ottersten. Automatic modulation classification for adaptive power control in cognitive satellite communications. In *2014 7th Advanced Satellite Multimedia Systems Conference and the 13th Signal Processing for Space Communications Workshop (ASMS/SPSC)*, pages 234–240. IEEE, 2014.

10 Milan Cabarkapa, Natasa Neskovic, Aleksandar Neskovic, and Djuradj Budimir. Adaptive nonlinearity compensation technique for 4G wireless transmitters. *Electron. Lett.*, 48(20):1, 2012.

11 Muhammad Asad Ullah, Konstantin Mikhaylov, and Hirley Alves. Analysis and simulation of LoRaWAN LR-FHSS for direct-to-satellite scenario. *IEEE Wirel. Commun. Lett.*, 11(3):548–552, 2021.

12 Rami Akeela and Behnam Dezfouli. Software-defined radios: Architecture, state-of-the-art, and challenges. *Comput. Commun.*, 128:106–125, 2018.

13 Manel Kammoun, Manel Elleuchi, Mohamed Abid, and Mohammed S BenSaleh. FPGA-based implementation of the SHA-256 hash algorithm. In *2020 IEEE International Conference on Design & Test of Integrated Micro & Nano-Systems (DTS)*, pages 1–6. IEEE, 2020.

14 Tenkasi V Ramabadran and Sunil S Gaitonde. A tutorial on CRC computations. *IEEE Micro.*, 8(4):62–75, 1988.

15 Peter M Chen, Edward K Lee, Garth A Gibson, Randy H Katz, and David A Patterson. RAID: High-performance, reliable secondary storage. *ACM Comput. Surv.*, 26(2):145–185, 1994.

16 Pingyue Yue, Jianping An, Jiankang Zhang, et al. Low earth orbit satellite security and reliability: Issues, solutions, and the road ahead. *IEEE Commun. Surv. Tutor.*, 25(3):1604–1652, 2023.

17 Yongtao Su, Yaoqi Liu, Yiqing Zhou, Jinhong Yuan, Huan Cao, and Jinglin Shi. Broadband LEO satellite communications: Architectures and key technologies. *IEEE Wirel. Commun.*, 26(2):55–61, 2019.

18 Alex A Kazemi and Allen Panahi. Space-based laser systems for inter-satellite communications. In *Photonic Applications for Aerospace, Transportation, and Harsh Environment III*, volume 8368, pages 156–164. SPIE, 2012.

19 Francesco Vatalaro, Giovanni Emanuele Corazza, Carlo Caini, and Carlo Ferrarelli. Analysis of LEO, MEO, and GEO global mobile satellite systems in the presence of interference and fading. *IEEE J. Sel. Areas Commun.*, 13(2):291–300, 1995.

20 Hiroki Nishiyama, Daigo Kudoh, Nei Kato, and Naoto Kadowaki. Load balancing and QoS provisioning based on congestion prediction for GEO/LEO hybrid satellite networks. *Proc. IEEE*, 99(11):1998–2007, 2011.

21 Willem Wijmans and Philippe Armbruster. Reconfigurable on-board payload data processing system developments at the European Space Agency. In *Imaging Spectrometry VI*, volume 4132, pages 91–98. SPIE, 2000.

22 Toshinori Kuwahara, Felix Böhringer, Albert Falke, Jens Eickhoff, Felix Huber, and Hans-Peter Röser. Fpga-based operational concept and payload data processing for the Flying Laptop satellite. *Acta Astronaut.*, 65(11–12):1616–1627, 2009.

23 Nedo Celandroni. Comparison of FEC types with regard to the efficiency of TCP connections over AWGN satellite channels. *IEEE Trans. Wirel. Commun.*, 5(7):1735–1745, 2006.

24 J Anzalchi, A Couchman, Piero Gabellini, et al. Beam hopping in multi-beam broadband satellite systems: System simulation and performance comparison with non-hopped systems. In *2010 5th Advanced Satellite Multimedia Systems Conference and the 11th Signal Processing for Space Communications Workshop*, pages 248–255. IEEE, 2010.

25 Jiajia Liu, Yongpeng Shi, Zubair Md Fadlullah, and Nei Kato. Space-air-ground integrated network: A survey. *IEEE Commun. Surv. Tutor.*, 20(4):2714–2741, 2018.

26 Deren Li, Qingxi Tong, Rongxing Li, Jianya Gong, and Liangpei Zhang. Current issues in high-resolution earth observation technology. *Sci. China Earth Sci.*, 55: 1043–1051, 2012.

27 Mélisande Cardona, Anton Schwarz, B Burcin Yurtoglu, and Christine Zulehner. Demand estimation and market definition for broadband internet services. *J. Regul. Econ.*, 35:70–95, 2009.

28 Chang Ge, Ning Wang, Ioannis Selinis, et al. QoE-assured live streaming via satellite backhaul in 5G networks. *IEEE Trans. Broadcast.*, 65(2):381–391, 2019.

29 Irfan Ahmed, Hedi Khammari, Adnan Shahid, et al. A survey on hybrid beamforming techniques in 5G: Architecture and system model perspectives. *IEEE Commun. Surv. Tutor.*, 20(4):3060–3097, 2018.

30 Taha Yasini, Jafar Roshanian, and Amir Taghavipour. Improving the low orbit satellite tracking ability using nonlinear model predictive controller and genetic algorithm. *Adv. Space Res.*, 71(6):2723–2732, 2023.

31 Leyuan Sun, Yueke Wang, Wende Huang, Jun Yang, Yifan Zhou, and Daoning Yang. Inter-satellite communication and ranging link assignment for navigation satellite systems. *GPS Solut.*, 22:1–14, 2018.

32 Heena Arora and Rakesh Goyal. A review on inter-satellite link in inter-satellite optical wireless communication. *J. Opt. Commun. Netw.*, 38(1):63–67, 2017.

33 Morio Toyoshima. Recent trends in space laser communications for small satellites and constellations. *J. Lightwave Technol.*, 39(3):693–699, 2021.

34 Martynas Milaševičius and Laurynas Mačiulis. A review of mechanical fine-pointing actuators for free-space optical communication. *Aerospace*, 11(1):5, 2023.

35 Phong Xuan Do and Shinichi Nakasuka. LEO-to-GEO laser communication-an enhancement of the communication capability of micro/nano-satellites. In *AIP Conference Proceedings*, volume 2366. AIP Publishing, 2021.

36 Biying Wang, Zheng Chang, Shancang Li, and Timo Hämäläinen. An efficient and privacy-preserving blockchain-based authentication scheme for low earth orbit satellite-assisted internet of things. *IEEE Trans. Aerosp. Electron. Syst.*, 58(6): 5153–5164, 2022.

37 Muhammad Baqer Mollah, Jun Zhao, Dusit Niyato, et al. Blockchain for the internet of vehicles towards intelligent transportation systems: A survey. *IEEE Internet Things J.*, 8(6):4157–4185, 2020.

38 Tianbo Lu, Xia Ding, Jiaze Shang, Pengfei Zhao, and Han Zhang. DoSat: A DDoS attack on the vulnerable time-varying topology of LEO satellite networks. In *International Conference on Applied Cryptography and Network Security*, pages 265–282. Springer, 2024.

39 Tudor Gabriel, Andrei Cornel-Cristian, Madalina Arhip-Calin, and Alexandru Zamfirescu. Cloud storage. A comparison between centralized solutions versus decentralized cloud storage solutions using blockchain technology. In *2019 54th International Universities Power Engineering Conference (UPEC)*, pages 1–5. IEEE, 2019.

40 Mauro Conti, Nicola Dragoni, and Viktor Lesyk. A survey of man in the middle attacks. *IEEE Commun. Surv. Tutor.*, 18(3):2027–2051, 2016.

41 Shan Kang, Naiwen Chen, Mi Yan, and Xiaoxiao Chen. Detecting identity-spoof attack based on BP network in cognitive radio network. In *Proceedings of 2011 Cross Strait Quad-Regional Radio Science and Wireless Technology Conference*, volume 2, pages 1603–1606. IEEE, 2011.

42 Adya Kiran, Samvid Dharanikota, and Annappa Basava. Blockchain based data access control using smart contracts. In *TENCON 2019-2019 IEEE Region 10 Conference (TENCON)*, pages 2335–2339. IEEE, 2019.

43 Ming Feng and Hao Xu. MSNET-blockchain: A new framework for securing mobile satellite communication network. In *2019 16th Annual IEEE International Conference on Sensing, Communication, and Networking (SECON)*, pages 1–9. IEEE, 2019.

44 Aleksandr Gelgor, Jon Montalbán, Dmitry Tkachenko, et al. Flexible satellite direct-to-home services with layered-division multiplexing. *IEEE Trans. Broadcast.*, 67(1):83–95, 2020.

45 Jie He, Wei Qiao, Jingyang Zhang, Lixiang Huang, and Lu Peng. Payload scheme research based on flexible access strategy of satellite constellation. In *2023 7th International Conference on Machine Vision and Information Technology (CMVIT)*, pages 134–138. IEEE, 2023.

46 Giannis Giannoulis, Aristeidis Stathis, Argiris Ntanos, et al. Satellite-to-ground QKD feasibility analysis for high altitude rural areas: A case study in Greece.

In *2024 International Workshop on Fiber Optics in Access Networks (FOAN)*, pages 38–42. IEEE, 2024.

47 Miralem Mehic, Marcin Niemiec, Stefan Rass, et al. Quantum key distribution: A networking perspective. *ACM Comput. Surv.*, 53(5):1–41, 2020.

48 Shengkai Liao, Wenqi Cai, Weiyue Liu, et al. Satellite-to-ground quantum key distribution. *Nature*, 549(7670):43–47, 2017.

49 Yu Fu and Xiufeng Zhao. Secure two-party dilithium signing protocol. In *2021 17th International Conference on Computational Intelligence and Security (CIS)*, pages 444–448. IEEE, 2021.

50 Pierre-Alain Fouque, Jeffrey Hoffstein, Paul Kirchner, et al. Falcon: Fast-Fourier lattice-based compact signatures over NTRU. *Submission to the NIST's post-quantum cryptography standardization process*, 36(5):1–75, 2018.

51 Daniel J Bernstein, Andreas Hülsing, Stefan Kölbl, Ruben Niederhagen, Joost Rijneveld, and Peter Schwabe. The sphincs+ signature framework. In *Proceedings of the 2019 ACM SIGSAC Conference on Computer and Communications Security*, pages 2129–2146, 2019.

52 Selvamani Chandrasekaran, K I Ramachandran, Sasidharan Adarsh, and Ashish Kumar Puranik. Avoidance of replay attack in CAN protocol using authenticated encryption. In *2020 11th International Conference on Computing, Communication and Networking Technologies (ICCCNT)*, pages 1–6. IEEE, 2020.

10

Joint Security and Reliability Design

10.1 Spaceborne Security

Space security can be articulated as the array of policies, technological innovations, and procedural frameworks devised to safeguard the integrity, confidentiality, and accessibility of extraterrestrial assets and their corresponding infrastructure. This paradigm encompasses a comprehensive spectrum of components, including satellites, communication pathways, and terrestrial stations, as well as the relevant hardware and software systems. The primary objective of space security is to guarantee that these systems sustain operational functionality in the face of unauthorized interference, data breaches, or physical impairments, which is especially vital in contexts that involve sensitive and critical applications.

In the domain of SCSs, the fundamental aspect of space security pertains to the safeguarding of the end-to-end communications continuum, extending from the orbiting satellite to the ground station and ultimately to the end user. This encompasses the protection of telemetry, tracking, and control (TT&C) data, alongside the assurance of secure uplink and downlink communications. The escalating number and intricacy of satellite systems, characterized by the prevalent deployment of LEO satellites, satellite constellations, and sophisticated interplanetary communication technologies, further accentuate the imperative of fortifying space security protocols.

Space security encompasses not solely the realm of physical protection but also the domain of cyber defense. While physical security aims to safeguard against kinetic threats, such as assaults on space assets via anti-satellite weaponry, cybersecurity is primarily concerned with mitigating vulnerabilities pertaining to data transmission, software integrity, and network architecture. As satellite systems increasingly converge with terrestrial networks, their susceptibility to prospective threats escalates. Consequently, space security has emerged as a paramount

Safeguarding Satellite Communications: Issues, Challenges, and Solutions, First Edition.
Jianping An, et al.

concern necessitating an integrated approach encompassing protective strategies across both physical and cyber realms.

10.1.1 Importance and Challenge of Spaceborne Security in SCSs

Space security is imperative for the dependable functioning of SCSs, which constitute the foundational infrastructure for numerous vital applications. These systems facilitate instantaneous data transmission, worldwide connectivity, and crucial services including navigation, terrestrial observation, and remote sensing. The significance of space security may be underscored in the subsequent domains:

- **Defense and National Security:** Satellites play a pivotal role in military communication, surveillance, and intelligence. A breach in satellite security could compromise national security, leading to significant operational and strategic risks.
- **Commercial Applications:** Industries such as telecommunications, broadcasting, and internet services heavily rely on satellites. Ensuring secure data transmission protects businesses from financial losses and reputational damage.
- **Global Communications:** Satellite systems provide connectivity to remote areas where terrestrial infrastructure is not feasible. Securing these systems ensures uninterrupted access to communication for billions of people worldwide.
- **Emergency Services:** In disaster scenarios, satellites facilitate coordination and communication for relief efforts. Secure systems are critical to ensuring accurate information flow during emergencies.
- **Space Exploration and Research:** Scientific missions depend on secure communication links to send and receive data. A breach in security could lead to the loss of mission-critical information and substantial financial setbacks.

Spaceborne security architecture: Space security encompasses a comprehensive array of technical and administrative strategies implemented to safeguard satellite systems and their communication links within both extraterrestrial and terrestrial contexts, as illustrated in Figure 10.1. This domain includes, though is not confined to, the safeguarding of the satellite itself, the encryption of communication channels, the fortification of signal integrity, the assurance of data confidentiality, and the protection of ground station facilities. In particular, spatial security encompasses the following critical dimensions:

- **Secure Satellite Payload Design:** The design of secure satellite payloads necessitates the establishment of resilience against a multitude of potential attacks while concurrently ensuring dependable and secure communication

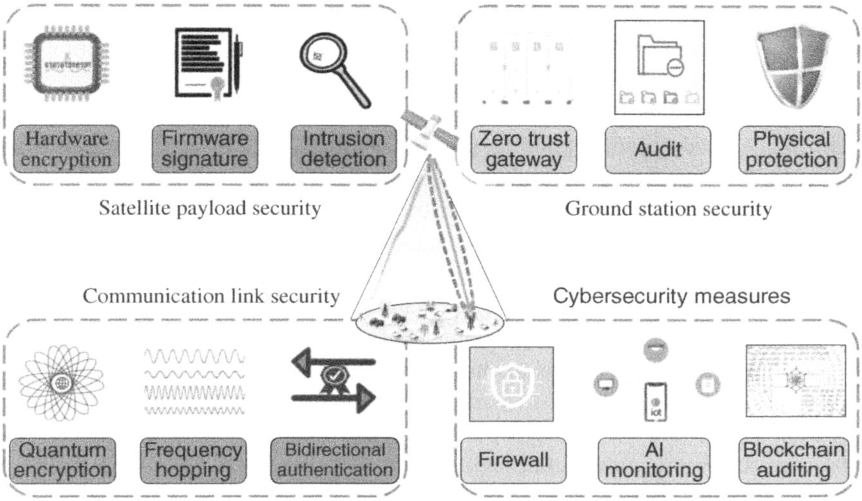

Figure 10.1 A schematic diagram of spaceborne security architecture.

channels. This process encompasses the implementation of stringent physical protection protocols, the prevention of hijacking or destruction of payloads, and the attenuation of the adverse effects arising from electromagnetic interference and electromagnetic pulse attacks.

- **Ground Station Security:** In order to facilitate secure communication links with terrestrial ground stations, it is imperative to thwart network intrusions and data breaches, while simultaneously safeguarding ground facilities from cyber threats and physical harm. Such measures are crucial for maintaining the integrity and security of communication commands and data exchanges.
- **Communication Link Security:** The application of encrypted communication links serves to ensure the security and reliability of data transmission between satellites and ground stations, effectively obstructing data interception or theft throughout the transmission process. Sophisticated encryption methodologies are employed to protect data integrity and confidentiality, thereby providing extensive safeguards against the potential leakage of information.
- **Cybersecurity Measures:** The implementation of network security protocols within SCSs, which includes firewalls, IDSs, and malware defense mechanisms, is essential for safeguarding the integrity of transmitted data. Furthermore, authentication protocols are utilized to ascertain the authenticity and reliability of communications.

Challenges: As previously articulated, space communications encounter a diverse array of security challenges that have the potential to undermine the integrity, confidentiality, and availability of communication systems. These challenges encompass, but are not restricted to, the following: malicious attacks, jamming, eavesdropping, spoofing, physical damage, data integrity and authentication, DoS attacks, geopolitical threats, emerging quantum and advanced cryptographic threats, and resource constraints.

Addressing these multifaceted challenges necessitates a holistic approach that incorporates strong encryption methodologies, resilient system architectures, real-time surveillance mechanisms, international cooperative efforts, and ongoing advancements in technology. By placing a premium on the security of space communication systems, relevant stakeholders can effectively mitigate potential risks and ensure the sustainability of operations within this vital sector. Consequently, the establishment of a standardized framework is imperative for the advancement of secure communication technologies.

10.1.2 Design Criteria for Spaceborne Security

The development of robust satellite communication architectures represents a complex undertaking that requires an extensive methodology to guarantee confidentiality, integrity, availability, and resilience in the face of diverse threats.

1. **Confidentiality and Encryption:** Confidentiality holds critical significance in the realm of satellite communications, functioning as the fundamental basis of information security. An array of encryption methodologies has been widely embraced, based on a collection of relevant standards. The ITU-T X.500 series standards highlight the importance of identity verification and information safeguarding within communication systems, thus establishing a foundational framework for the deployment of encryption technologies [1].

   NIST SP 800-53 Rev. 5 articulates comprehensive protocols for data encryption and transmission security control measures, thereby functioning as a reference framework for the systematic implementation of encryption technologies [2]. In a similar vein, NIST SP 800-171 Rev. 2 emphasizes the encryption security of unclassified information within federal systems, thereby safeguarding the confidentiality of data across diverse communication contexts [3].

   In the realm of practical technical applications, the AES-256 encryption algorithm has emerged as a fundamental technology for facilitating secure communication between satellites and terrestrial stations [4]. With a key length of 256 bits, this algorithm proficiently and dependably encrypts data disseminated over satellite communication channels, encompassing remotely sensed data transmitted on the uplink and control commands relayed on the downlink. Through this encryption mechanism, the original information is converted into an incomprehensible ciphertext, which can solely be deciphered by authorized entities in possession of the corresponding 256-bit key. This procedure effectively thwarts unauthorized access and interception of data during transmission, thus preserving the confidentiality of satellite communications.

2. **Integrity and Authentication:** In SCSs, the preservation of data integrity and the validation of data sources represent pivotal measures in the enhancement of communication security. This procedure rigorously complies with an array of international standards and incorporates sophisticated technical methodologies to fulfill these objectives. The ITU-T X.509 series standards provide a substantial foundational paradigm for the implementation of digital signatures and certificate validation. Through the utilization of digital certificates, these standards ascertain the authentication of the sender's identity, thereby ensuring the dependability and traceability of digital signatures [5]. The ETSI EN 303 413 standard is specifically directed towards the security of satellite ground station communications, with its authentication component acting as an essential safeguard against unauthorized entry and ensuring secure data transmission [6]. NIST SP 800-53 Rev. 5 delineates comprehensive definitions of integrity protection and authentication mechanisms, detailing the technical implementation processes and methodologies [2].

   In practical applications, the SHA-256 hash algorithm, in conjunction with digital signatures, constitutes a fundamental approach for the verification of data integrity. The sender generates a digital signature utilizing their private key alongside the SHA-256 hash of the data. As a secure hash algorithm, SHA-256 is capable of processing data of any length to yield a unique 256-bit hash value, which serves as the "fingerprint" of the data. Even minimal modifications to the data result in a substantially different hash value. Upon receipt of the data and its corresponding digital signature, the recipient initially verifies the signature using the sender's public key. Subsequently, the recipient recalculates the SHA-256 hash of the received data and contrasts the resultant hash with the hash transmitted alongside the data. A precise correspondence between the two hashes strongly suggests that the data has remained unaltered during transmission, while the authenticated digital signature ascertains the reliability of the data source.

   Moreover, multifactor authentication mechanisms are extensively employed to bolster security measures. These mechanisms amalgamate biometric technologies (such as fingerprint recognition and facial recognition), smart card authentication, and conventional username/password authentication methods. For example, fingerprint recognition exhibits an exceptionally low error rate, generally below 0.01%, significantly enhancing the precision and security of identity verification. By integrating multiple authentication techniques, a comprehensive and multilayered defense is established for the identity verification processes within SCSs. This effectively obstructs unauthorized users from gaining access to the system and ensures the integrity and security of communication data.

3. **Availability and Redundancy:** In SCSs, the assurance of availability and redundancy is of paramount importance for the sustenance of uninterrupted and stable communication services. The realization of this objective necessitates rigorous compliance with pertinent standards and the implementation of sophisticated technical architectures.

   The ETSI EN 303 413 standard delineates explicit requirements for redundancy design within SCSs operating on mobile platforms [6]. This standard underscores the intricate and dynamic contexts in which mobile platforms function, necessitating that SCSs are endowed with redundancy capabilities to mitigate potential signal disruptions and ensure the continuity of communication during mobility. In a parallel vein, ETSI EN 302 307 concentrates on redundancy mechanisms pertinent to ground stations and communication systems, encompassing aspects such as equipment redundancy, link redundancy, and other relevant considerations [7]. Its overarching aim is to diminish the likelihood of system failures precipitated by singular points of failure.

   Moreover, ITU-R S.1503 strives to preserve the stability of ground station antennas by stipulating regulations concerning their structural design, installation methodologies, and protective strategies [8]. This regulation guarantees that antennas retain operational reliability even amidst adverse meteorological conditions or electromagnetic interference, thereby facilitating stable ground-based reception and transmission support for satellite communication endeavors. Furthermore, FCC 47 CFR Part 25 imposes a mandate that SCSs must exhibit interference resilience and redundancy features, instituting regulatory frameworks to ensure the operational availability of SCSs within intricate electromagnetic settings.

   With respect to technological implementations, the integration of multiband communication links serves as a pivotal strategy for augmenting link availability. For instance, the simultaneous utilization of communication across the C-band (3.7–4.2 GHz), Ku-band (11.7–12.2 GHz), and Ka-band (27.5–30 GHz) capitalizes on the distinct propagation characteristics and interference mitigation capabilities inherent to each band. In instances where one band encounters disruption, the system is capable of autonomously transitioning to an alternative band, thereby guaranteeing the continuity of communication.

The design of backup satellites constitutes another essential strategy for preserving the availability of communication links. By deploying a constellation of backup satellites that operate in conjunction with primary satellites, the system can execute a seamless transition to a backup satellite in the event of primary satellite failure, hostile actions, or other operational anomalies. Backup satellites typically remain in a hot standby state, perpetually monitoring the operational status of the primary satellite. Upon the detection of irregularities, these satellites can effectuate the switchover process within a remarkably brief timeframe, thereby ensuring the continuity of communication services.

The synthesis of multi-band communication links and backup satellite designs furnishes robust support for the availability and redundancy of SCSs. This comprehensive strategy empowers the system to function reliably within complex and variable spatial environments and communication conditions, thereby delivering consistent and dependable communication services to end users.

4. **Access Control:** In the domain of satellite communications, access control is an essential element for the enhancement of system security and the deterrence of unauthorized access. Its execution is predicated upon rigorous compliance with pertinent standards and the application of sophisticated technical methodologies.

   The FCC Title 47 CFR Part 25 delineates explicit regulatory frameworks for the secure acquisition of communication systems and user authentication protocols. These statutory provisions guarantee that access to SCSs is restricted solely to individuals with proper authorization, thereby establishing a fundamental legal foundation for access control mechanisms.

   ETSI TS 102 900 pertains to digital broadcasting systems, delineating standards for authentication and secure access [9]. Through the implementation of robust authentication mechanisms, it guarantees that only duly authorized individuals are able to receive and process digital broadcast signals, thereby fortifying the security of satellite communications within the broadcasting sector. Furthermore, the ITU-T X.500 series underpins directory service security, thereby facilitating effective role-based access control (RBAC). The RBAC paradigm allocates access permissions predicated on the roles of users within an organization, thereby significantly streamlining permission management while concurrently enhancing both efficiency and security.

   From a technological standpoint, the implementation of biometric technologies has markedly enhanced the precision and security of user authentication processes. For instance, fingerprint recognition technology, which typically exhibits a false acceptance rate of less than 0.01%, proficiently identifies users and effectively mitigates the risk of identity impersonation.

MFA further fortifies access security by amalgamating biometrics, smart card authentication, and conventional password validation. In order to successfully complete the authentication process, users are required to furnish biometric data, such as fingerprints, insert an appropriate smart card, and input the correct password. This considerably heightens the complexity for unauthorized individuals attempting to circumvent the system.

Role-based permission assignment frameworks also significantly contribute to access control by meticulously delineating access rights in accordance with users' roles and responsibilities. For example, satellite system administrators are endowed with the highest tier of permissions, thereby enabling them to configure and modify various satellite parameters. Conversely, ordinary data analysts are confined to read-only access to designated databases and are prohibited from executing any alterations. This methodology ensures that each user operates exclusively within their authorized parameters, effectively thwarting unauthorized access and reducing the risk of data breaches.

Through the integration of these standards, technologies, and management practices, SCSs attain a comprehensive level of access security, thereby guaranteeing robust protection against unauthorized access and preserving the integrity of operational processes.

5. **Anti-jamming and Anti-spoofing:** ITU-R M.1583-2 delineates a methodologically rigorous and scientifically grounded framework for the assessment of frequency interference in SCSs. By employing this standard, engineers are afforded the capacity to perform comprehensive analyses and forecasts regarding potential interference across diverse frequency bands, thereby facilitating the development of preemptive strategies aimed at alleviating interference and minimizing its detrimental effects on communication infrastructures [10]. In a parallel vein, ITU-R S.1003-5 is dedicated to the characterization of antenna radiation patterns. By attaining meticulous control over the radiative properties of antennas, this standard proficiently mitigates the reception of extraneous interfering signals. The judicious design of antenna radiation profiles guarantees the reception of pertinent signals while concurrently curbing the acquisition of interference, thereby furnishing essential hardware support for anti-interference initiatives.

   From the standpoint of system design, the ETSI EN 302 307 standard codifies anti-interference protocols. It necessitates that SCS architectures integrate considerations regarding potential interference sources and implement specific protective strategies [7]. For instance, the optimization of circuit design and signal transmission pathways can diminish internal interference while bolstering resilience against external disruptions. ESA ECSS-E-ST-10-03C advocates for a holistic systems engineering approach to counteracting interference and spoofing [11]. This standard conceptualizes the SCS as an integrated entity, encompassing all components from the satellite transmission point to the ground reception terminal. By embedding anti-interference and anti-spoofing measures at each developmental phase and utilizing coordinated design and optimization at the system level, it substantially enhances the overall performance of the system in combating interference and spoofing attempts.

In terms of technical applications, frequency-hopping technology assumes a pivotal role. By rapidly transitioning among multiple frequency points according to a designated PR sequence, this technique complicates the ability of interference sources to synchronize with the satellite communication frequency, thereby significantly mitigating the adverse impacts. Waveform pseudo-randomization further complicates satellite signal waveforms through random modulation, thereby augmenting signal stability in the presence of interference. Adaptive antenna technology dynamically modifies the radiation pattern of the antenna in response to the direction and intensity of interfering signals. Upon detection of interference, the antenna can judiciously adjust its beam orientation to circumvent the interference source, thereby enhancing its capacity to receive valuable signals.

Moreover, BCH coding, a sophisticated error-correction methodology, is instrumental in reinforcing anti-interference and anti-spoofing strategies. By incorporating redundant information into transmitted signals, BCH coding empowers the receiver to identify and rectify errors induced by interference during the transmission process. Its distinctive characteristics also render it remarkably resilient to spoofing attempts, as counterfeit signals are unlikely to replicate the precise coding attributes of authentic signals.

Through the synthesis of these advanced technologies, the anti-interference and anti-spoofing capabilities of SCSs are substantially augmented, thereby ensuring the dependability and security of communication services.

10.1.3 Regulations and Standards for Spaceborne Security

International regulations: Regulations and legal frameworks at the international level establish the foundational structure and guidance necessary for the safeguarding of global SCSs. Prominent international entities, including the ITU, the United Nations Office for Outer Space Affairs (UNOOSA), and various international cooperative organizations, are instrumental in the formulation and dissemination of pertinent safety standards and regulations. The subsequent discussion delineates some of the pivotal international regulations and policies:

ITU: In the domain of space security, the ITU assumes a vital role through its Radio Regulations, which delineate mechanisms for the governance and oversight of satellite communications. These regulations are designed to ensure that the utilization of RF spectrum on a global scale does not engender conflicts, while concurrently imposing stipulations concerning satellite system launches, orbital allocations, and the security of satellite communications. More specifically, ITU's RF management rules not only delineate the allocations of SCSs within the spectrum but also articulate procedures for the resolution of interference issues. To alleviate disruptions in communication links, ITU advocates for enhanced collaboration among member states, thereby bolstering the stability and security of such links. Additionally, ITU has formulated standardized frameworks aimed at ensuring the security of SCSs, particularly with respect to extensive satellite constellations such as LEO networks. Through the dissemination of guidelines and best practices, ITU endeavors to mitigate risks associated with spectrum conflicts, communication interference, and unauthorized interception.

UNOOSA: The UNOOSA is committed to fostering international collaboration and the peaceful utilization of outer space. By publishing the Outer Space Treaty and other pertinent documents, UNOOSA enhances the security of global satellite communications. The Outer Space Treaty, instituted in 1967, delineates fundamental principles governing space exploration and satellite launches, with a strong emphasis on the peaceful use of outer space and the prohibition of its militarization. Furthermore, the treaty offers guidelines aimed at ensuring security in space activities, particularly in the prevention of interference, interception, or assaults on satellite communications. It explicitly mandates that space activities undertaken by any nation must not impede the space assets of other states.

10.2 Spaceborne Reliability

Spaceborne security has emerged as an indispensable element of contemporary information ecosystems and defense infrastructures, propelled by the increasing dependence on satellite systems for functions such as communication, navigation, reconnaissance, and surveillance. In contrast to terrestrial systems, satellites encounter distinct challenges, including extreme radiation exposure, thermal cycling, vacuum conditions, and cyber-physical vulnerabilities. Given the inherent difficulties associated with performing routine maintenance or human intervention once spaceborne systems are operational, it is imperative to ensure the stable and reliable functioning of these systems to secure the success of mission-critical operations amidst various threats, thereby underscoring the significance of spaceborne reliability.

Spaceborne reliability pertains to a system's capacity to sustain its intended performance throughout its operational lifespan, even when exposed to the harsh and dynamic conditions prevalent in outer space. Unlike conventional systems that can be serviced or repaired as challenges emerge, spaceborne systems must function autonomously without direct physical intervention following deployment. Consequently, reliability plays a pivotal role in determining mission success and serves as a fundamental criterion for assessing system performance. Elevated reliability not only guarantees that the system consistently achieves its objectives but also mitigates mission risks and the costs associated with unforeseen failures. Within the framework of spaceborne security, reliability assumes particular importance, as any unanticipated downtime or malfunction could engender exploitable vulnerabilities, jeopardize sensitive data, or disrupt crucial operations.

10.2.1 Introduction to Spaceborne Reliability

In the conceptualization and functionality of extraterrestrial systems, the assurance of reliability assumes an imperative significance owing to the severe and fluctuating circumstances that these systems encounter in the cosmos. The reliability of spaceborne systems is conventionally categorized into four principal domains: hardware reliability, software reliability, mission reliability, and redundancy coupled with fault tolerance, as illustrated in Figure 10.2. Each of these domains serves a crucial function in the operational efficacy of satellites, thereby facilitating spaceborne systems to sustain optimal performance throughout prolonged missions, notwithstanding the demanding and unpredictable environments inherent to outer space.

- **Hardware Reliability:** Hardware reliability pertains to the resilience and longevity of the tangible components of a satellite, encompassing electronic circuits, solar panels, antennas, and propulsion mechanisms. Spaceborne hardware is required to withstand extreme environmental conditions, which include severe radiation exposure, thermal cycling, and the potential impact of high-velocity extraterrestrial debris. Moreover, the assurance of hardware reliability is particularly problematic due to the irreparable nature of systems deployed in space.
- **Software Reliability:** Software reliability is fundamental in ensuring the trustworthiness of onboard software, which is pivotal in sustaining system stability. The implementation of fault-tolerant algorithms, robust encryption methods, and dependable communication protocols is imperative to avert system failures arising from unforeseen inputs or environmental perturbations. Mechanisms for error detection and correction, when integrated with autonomous fault recovery systems, serve to alleviate potential disruptions and avert transient anomalies from escalating into critical mission failures.

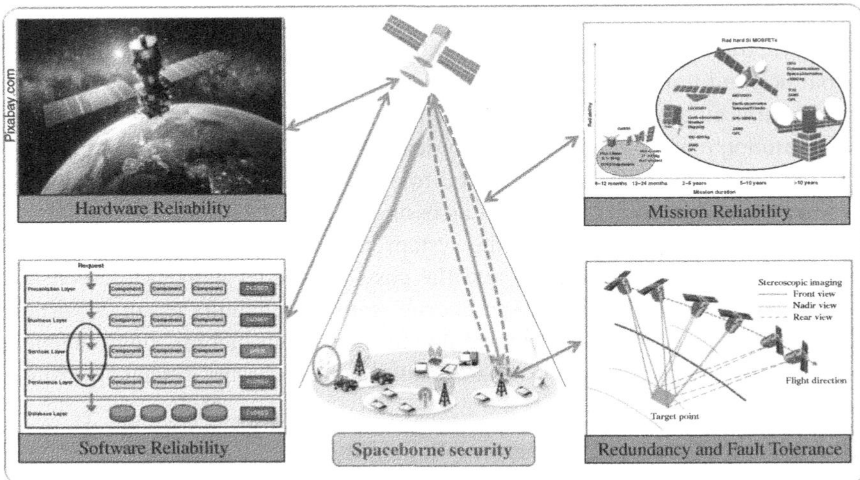

Figure 10.2 A schematic diagram of spaceborne reliability architecture.

- **Mission Reliability:** Mission reliability emphasizes the comprehensive performance of the satellite in fulfilling its designated mission objectives. A fundamental aspect is the assurance of communication reliability, which facilitates the uninterrupted functionality of spaceborne systems. The fidelity of communication links, minimal latency, and adequate throughput are essential to guaranteeing the unobstructed transmission of commands, telemetry, and data over extensive distances. These communication links must maintain stability and robustness to enable real-time control, situational awareness, and prompt decision-making.
- **Redundancy and Fault Tolerance:** Redundancy and Fault Tolerance constitute essential methodologies for augmenting reliability in spaceborne systems. Redundant architectures, including auxiliary power supplies or alternative communication pathways, are implemented to guarantee that, in the occurrence of a malfunction, a secondary component can swiftly assume functionality, thereby mitigating service interruptions and minimizing periods of inactivity. The implementation of cross-strapping, which entails the interconnection of redundant subsystems for the purpose of load-sharing, further ensures that failures do not propagate throughout the system. These methodologies enhance the fault tolerance of spaceborne systems, thereby bolstering mission reliability by offering supplementary options and diminishing the probability of catastrophic failures.

The formidable environment of outer space poses considerable challenges to the assurance of reliability in spaceborne systems. In the absence of opportunities for direct maintenance, meticulous pre-launch testing and stringent quality assurance protocols are imperative to avert critical failures. Prior to deployment, spaceborne systems are subjected to an extensive array of tests designed to replicate the extreme conditions encountered in space. These rigorous evaluations facilitate the identification and rectification of potential vulnerabilities, thereby mitigating the risk of failure subsequent to launch. Quality assurance mechanisms further guarantee that each component operates as designed, thereby equipping the satellite to endure the demanding environmental and operational challenges inherent to its mission.

Historically, the discipline of satellite reliability engineering has regarded security and reliability as distinct domains. Nevertheless, the escalating incidence of cyberattacks targeting space infrastructure underscores the necessity for a cohesive strategy to protect mission continuity from both natural disturbances and adversarial threats. For instance, hardware failures induced by radiation could inadvertently reveal security vulnerabilities, while excessively intricate encryption protocols might burden OBP resources, detrimentally affecting system reliability. As contemporary spaceborne systems continue to scale, the management of reliability becomes increasingly intricate. Moreover, the escalating cybersecurity threats necessitate the implementation of robust encryption and secure communication protocols. However, the attainment of high reliability must also be judiciously balanced against cost limitations, necessitating careful trade-offs in engineering decisions to ensure enduring operational success.

10.2.2 Design Criteria for Spaceborne Reliability

The design criteria for spaceborne reliability encompass a comprehensive array of engineering principles and methodologies that are geared towards improving system robustness, fault tolerance, and operational longevity in the face of extreme conditions such as radiation, vacuum, and thermal fluctuations. A primary emphasis is placed on ensuring resilience against environmental stressors while simultaneously providing protection against cyber threats that could jeopardize system integrity. In addition, the incorporation of onboard decision-making capabilities is imperative for effective fault recovery, enabling spacecraft to autonomously identify and rectify anomalies without necessitating immediate intervention from ground control, thereby bolstering mission resilience. Another fundamental dimension is dynamic resource management, which facilitates the efficient allocation of resources to sustain an equilibrium between security measures, such as encryption processing, and the overall QoS provided.

Moreover, redundancy and fault tolerance constitute critical elements in guaranteeing uninterrupted operation, integrating backup systems and error-detection methodologies to alleviate the impact of failures.

10.2.2.1 Designing for Hardware Reliability

The hardware utilized in extraterrestrial environments serves as the essential cornerstone of satellite systems. In the realm of outer space, satellites encounter a multitude of severe conditions; therefore, the design methodology must prioritize the selection of materials and components that exhibit both durability and dependability in response to these exigent circumstances.

1. **Radiation-hardened Electronics**
   One of the foremost considerations in the reliability of spaceborne hardware pertains to radiation hardening. The space environment subjects electronic components to ionizing radiation, which can result in SEUs, latch-ups, and prolonged degradation. To alleviate these hazards, components that are either radiation-hardened or radiation-tolerant are meticulously selected.
   In the context of analog circuit applications, specialized radiation-hardened cells are utilized to guarantee stable performance. Design methodologies that exploit shallow trench isolation (STI) and silicon-on-insulator (SOI) technologies effectively diminish radiation-induced leakage currents and shifts in threshold voltage [12]. Radiation-hardened, fully depleted, SOI static random-access memories (SRAM), which integrate highly reliable elements and sophisticated power management circuits, exhibit substantial resilience against total ionizing dose (TID) and single event effects (SEEs), thereby ensuring elevated reliability in space applications [13]. The European Cooperation for Space Standardization (ECSS) delineates stringent qualification criteria for radiation-hardened components, encompassing TID limits (e.g., 100 krad for LEO missions) and SEE tolerance (e.g., 10–100 MeV/cm^2 for proton impact), thereby establishing a framework to ensure that components fulfill the requisite resilience for space operations [14].
2. **Thermal Management**
   Thermal cycling and exposure to extreme temperatures in the extraterrestrial environment—spanning from −150 °C in the Earth's shadow to +150 °C under direct solar irradiation—necessitate the implementation of efficient thermal control systems. The integration of multilayer insulation (MLI), heat pipes, and radiators is instrumental in sustaining operational temperature thresholds for delicate electronics and payloads. A notable instance is the Thermal Management System of the Mars Rover, which employs both active and passive mechanisms to ensure that its components remain within optimal operational temperature ranges [15].

The selection of materials is also of paramount importance. For instance, materials exhibiting high-temperature resistance, such as ceramic composites utilized for thermal shields and aluminum alloys employed in satellite structures, are strategically selected to mitigate degradation resulting from extended exposure to extreme temperatures. Furthermore, the outgassing of materials in the vacuum of space can present a considerable challenge, thereby necessitating the adoption of low-outgassing materials to avert contamination of optical and electronic systems.

3. **Mechanical Integrity**
   During the launch phase, spaceborne apparatus encounters substantial acceleration forces, vibratory motions, and acoustic pressures. It is imperative that components are engineered to withstand these physical stresses without experiencing failure. For instance, redundant actuators within ACS are frequently utilized to guarantee that the spacecraft can sustain appropriate orientation, notwithstanding the potential failure of a single actuator [16]. Launch vibration assessments and shock evaluations in accordance with NASA-STD-5001 are conducted to validate that structural components are capable of enduring the conditions encountered during launch [17].

10.2.2.2 Designing for Software Reliability

Spaceborne systems are fundamentally dependent upon onboard software to facilitate mission operations, fault detection, and subsequent fault recovery. The reliability of software is contingent upon its fault tolerance capabilities, autonomous recovery mechanisms, and cybersecurity protocols. The design process aimed at achieving software reliability necessitates the incorporation of fault-tolerant algorithms, redundancy strategies, and comprehensive security measures to ensure the continuation of operations in the face of unforeseen disturbances.

1. **Error Detection and Correction**
   The extraterrestrial environment is inherently susceptible to erratic phenomena, including cosmic radiation and solar flares, which may precipitate inaccuracies in data transmission. ECCs, such as Hamming codes, RS codes, and Turbo codes, are employed to identify and rectify inaccuracies in both data storage and transmission processes. For example, FEC is frequently utilized in communication pathways to alleviate the adverse effects of signal deterioration attributable to space weather occurrences. Model-based system engineering (MBSE) frameworks, as delineated in ECSS-E-ST-10-03, employ systems modeling language (SysML) to establish hierarchical reliability frameworks, thereby facilitating iterative validation of fault detection algorithms [18].

2. **Autonomous Fault Detection and Recovery**
   Spaceborne software must possess the capability to autonomously identify faults within the hardware and to initiate appropriate corrective measures. The autonomous fault management system (AFMS) utilized in the Hubble Space Telescope exemplifies how software can oversee satellite subsystems and initiate recovery actions, such as system rebooting or transitioning to redundant hardware in the event of a fault occurrence. The Rosetta mission serves as a pertinent illustration of the onboard software's proficiency in autonomously modifying trajectory and operational parameters in response to unforeseen failures.

   The ISO 16615 standard delineates the criteria for the stable orbital operation of the satellite attitude and orbit control system, ensuring that the spacecraft maintains a secure operational status under all conditions, including emergency situations [19].

10.2.2.3 Designing for Mission Reliability

Mission reliability pertains to the comprehensive performance of the satellite system, guaranteeing sustained, high-efficiency operation throughout its operational lifespan. An essential element of mission reliability is the accessibility and stability of communication links. Satellites depend on secure and continuous communication with terrestrial stations, necessitating that the system be engineered to mitigate risks such as solar flares, RF interference, or atmospheric attenuation. The Starlink LEO satellite constellation utilizes ISLs to augment network performance, which enables direct communication between satellites, thereby diminishing dependence on ground stations and consequently reducing latency [20]. The incorporation of laser communication technology within satellite networks is also recognized to enhance data transmission rates significantly. For instance, Amazon's Project Kuiper has reported that its prototype satellites, outfitted with infrared lasers, attained data transfer rates of 100 Gbps. This exemplifies the potential for high-throughput capabilities in contemporary satellite constellations that employ laser communication.

10.2.2.4 Designing for Redundancy and Fault Tolerance

In the inhospitable environment of outer space, wherein repairs are unfeasible, the implementation of redundancy and fault tolerance emerges as a crucial strategic imperative. Essential systems, including power supplies, communication frameworks, and computational hardware, must be replicated to guarantee that auxiliary systems can promptly assume control in the event of a primary system failure.

Redundancy is generally accomplished through two methodologies: cold redundancy and hot redundancy. Cold redundancy pertains to the maintenance of backup systems in an offline state until their utilization becomes necessary, whereas hot redundancy guarantees that backup systems operate concurrently with primary systems, thereby facilitating an instantaneous transition upon a failure. An illustration of cold redundancy is observable in the Mars Science Laboratory (MSL) mission, wherein backup systems for the rover's navigation apparatus remained inactive until a defect in the primary system manifested. Conversely, hot redundancy is exemplified by the Hubble Space Telescope, where critical components, such as gyroscopes, function through dual systems operating simultaneously.

Cross-strapping, defined as the interconnection of redundant subsystems for the purpose of load-sharing and failure mitigation, constitutes a fundamental design consideration. For instance, in SCSs, redundant communication channels are interconnected to ensure uninterrupted data transmission in the occurrence of a malfunction in one communication channel.

NASA has formulated internal guidelines that concentrate on fault-tolerant design, encompassing analytical design disciplines, failure modes, effects, and criticality analysis (FMECA), as well as fault tree analysis (FTA) investigations [21].

10.2.2.5 Integrated Design Approach for Reliability and Security

In contemporary space systems, the design of hardware, software, mission reliability, and security cannot be executed in a compartmentalized manner. A comprehensive design methodology that takes into account all dimensions of reliability and security in an integrated fashion is imperative to guarantee the efficacy of space-borne operations. For example, radiation-hardened hardware necessitates the augmentation of secure software protocols and autonomous fault recovery systems to prevent a malfunction in one component from jeopardizing the overall mission.

The amalgamation of security with reliability design is paramount in space systems, as weaknesses in a single domain can precipitate a series of failures. Encryption, intrusion detection, and fault tolerance must be cohesively incorporated to safeguard both the confidentiality and continuity of satellite operations. Through the integration of hardware redundancy, software resilience, mission-critical communication, and system security, spaceborne systems can attain the elevated levels of reliability required to confront the challenges associated with space exploration and operations.

A codesign framework harmonizes hardware durability, software resilience, and mission requirements. For instance, ESA's Galileo constellation exemplifies the integration of TMR-based atomic clocks with quantum-resistant encryption, thereby ensuring that navigation signals remain synchronized despite radiation-induced clock drifts. Furthermore, the design must proactively identify potential security threats that could compromise the system's reliability, such as the exploitation of vulnerabilities in mission-critical software or communication pathways. By embracing an integrated design philosophy, space missions can mitigate the probability of failure stemming from technical challenges or security infringements, ultimately reinforcing the system's resilience against both natural and anthropogenic risks.

10.2.3 Challenges and Future Directions of Satellite Reliability

Ensuring the reliability of spaceborne systems presents a multifaceted array of challenges attributable to the distinctive operational conditions these systems encounter. The space environment is fundamentally inhospitable, defined by severe radiation exposure, impacts from micrometeoroids, extreme variations in temperature, and the lack of an atmospheric shield. Such extreme environmental conditions exert considerable stress on spaceborne systems, thereby necessitating the formulation of designs that are exceptionally robust and resilient. Furthermore, as satellite systems increase in complexity, the management of reliability becomes progressively challenging. Contemporary satellites frequently amalgamate a diverse array of sensors, processors, and communication links. The interdependence of these components implies that a failure in any singular element of the system could precipitate a cascading effect throughout the entire network, particularly in expansive constellations. Consequently, comprehensive reliability evaluations at the system-wide level are imperative to identify potential vulnerabilities and mitigate the risks associated with failure. This complexity is further compounded by the increasing incorporation of digital communication systems, which render spaceborne systems susceptible to cybersecurity threats, including software vulnerabilities, communication jamming, and attempts to commandeer control. To counteract these risks, it is essential to implement robust encryption protocols, secure communication mechanisms, and effective IDS to preserve system integrity and ensure uninterrupted, reliable performance.

Looking ahead, several emergent challenges warrant consideration in the pursuit of ensuring spaceborne reliability. One of the most urgent issues is the escalation of cybersecurity threats, particularly the potential risks introduced by quantum computing; thus, the advancement of quantum-resistant encryption technologies will be crucial in preserving the security and reliability of spaceborne systems within this evolving paradigm. Moreover, the utilization of AI is becoming increasingly pivotal in the enhancement of spaceborne reliability. AI-driven reliability solutions possess the capacity to augment operational efficiency and diminish the necessity for costly interventions, thereby bolstering the overall resilience of spaceborne systems.

References

1 ITU-T. Information technology—open systems interconnection—the directory: Overview of concepts, models and services. Oct. 2019.

2 NIST. Security and privacy controls for information systems and organizations. Technical Report NIST Special Publication (SP) 800-53, Rev. 5, U.S. Department of Commerce, Gaithersburg, MD, 2020. Includes updates as of December 10, 2020.

3 NIST. Protecting controlled unclassified information in nonfederal systems and organizations. Technical Report NIST Special Publication 800-171, Revision 2, U.S. Department of Commerce, Gaithersburg, MD, January 2021. Superseded by Revision 3 in May 2024.

4 M Santhanalakshmi, M Lakshana, and GM Shahitya. Enhanced AES-256 cipher round algorithm for IoT applications. *Sci. Temper*, 14(01):184–190, 2023.

5 ITU. The directory authentication framework. Technical report, n.d.

6 ETSI. EN 303 413 v1.2.1: Satellite earth stations and systems (SES); global navigation satellite system (GNSS) receivers; GNSS user equipment (GUE). Technical report, 2021. Version 1.2.1.

7 ETSI. En 302 307 v1.2.1: Digital video broadcasting (DVB); second generation framing structure, channel coding and modulation systems for broadcasting, interactive services, news gathering and other broadband satellite applications. Technical report, 2009. Version 1.2.1.

8 ITU. Functional description to be used in developing software tools for determining conformity of non-geostationary-satellite orbit fixed-satellite service systems or networks with limits contained in Article 22 of the Radio Regulations. Technical report, 2023.

9 ETSI. TS 102 900 v1.3.1: European public warning service (EU-ALERT) using cell broadcast service (CBS). Technical report, ETSI, February 2019. Version 1.3.1.

10 ITU. Interference calculations between non-geostationary mobile-satellite service or radionavigation-satellite service systems and radio astronomy stations. Technical report, October 2007. ITU-R M.1583-2.

11 ECSS. ECSS-E-ST-10-03 rev.1: Space engineering—testing. Technical report, May 2022. Replaces ECSS-E-ST-10-03C.

12 K Vasilikou, Vasiliki Gogolou, and Thomas Noulis. Radiation-hardened cells for analog design: Comprehensive review and design flow. In *2024 13th International Conference on Modern Circuits and Systems Technologies (MOCAST)*, pages 1–4, 2024.

13 Chang Cai, Minchi Hu, Lei Shen, Jun Yu, and Gengsheng Chen. Design and evaluation of a radiation-hardened FDSOI SRAM with high-reliable elements and power management circuits for space application. *IEEE Trans. Aerosp. Electron. Syst.*, 61(2):5057–5066, Apr. 2025.

14 ECSS. Space product assurance—commercial electrical, electronic and electromechanical (EEE) components. Technical Report ECSS-Q-ST-60-13C Rev.1, ESA, Noordwijk, the Netherlands, May 2022.

15 Gaetano Quattrocchi, Andrea Pittari, Matteo Dalla Vedova, and Paolo Maggiore. The thermal control system of NASA's curiosity rover: A case study. In *IOP Conference Series: Materials Science and Engineering*, volume 1226, Feb. 2022.

16 Brian Kawauchi. Redundancy management for autonomous spacecraft attitude control systems. In *1986 American Control Conference*, pages 497–506, 1986.

17 NASA. Structural design and test factors of safety for spaceflight hardware. Technical Report NASA-STD-5001B w/Change 3, NASA, Washington, D.C., Oct. 2022.

18 ECSS. Space engineering—testing. Technical Report ECSS-E-ST-10-03C Rev.1, ECSS Secretariat, ESA-ESTEC, Noordwijk, the Netherlands, May 2022.

19 IOS. Space systems—stable operation requirements for satellite attitude and orbit control system. Technical Report ISO 1661 5: 2022, International Organization for Standardization, Geneva, Switzerland, 2022.

20 Aizaz U Chaudhry and Halim Yanikomeroglu. Temporary laser inter-satellite links in free-space optical satellite networks. *IEEE Open J. Commun. Soc.*, 3:1413–1427, 2022.

21 NASA. Fault management handbook. Technical Report NASA-HDBK-1002, NASA, Washington, D.C., Apr. 2012.

Part V

Future Directions

11

Secure Space Transmission Techniques

The militarization of outer space, the increasing congestion in orbital pathways, and the advancements in cyberattack methodologies have rendered satellites increasingly susceptible to electronic disruption, uncertainties in situational awareness, and threats to data security. Current technological frameworks exhibit limitations in satisfying the exigencies of real-time sensing, intelligent decision-making, and adaptable secure communication within satellite systems, thereby necessitating the incorporation of novel technologies to further augment satellite security and reliability. In response to these vulnerabilities, a variety of innovative strategies have emerged, encompassing ISAC, CV, multi-satellite secure transmission methodologies, and security-centric RF techniques, which present new avenues for the establishment of secure satellite communication.

This chapter will undertake a comprehensive examination of these sophisticated methodologies. ISAC augments the security of satellite communication networks by amalgamating sensing capabilities with communication technologies, thereby enhancing situational awareness and improving threat detection mechanisms. CV-aided communication employs CV techniques to facilitate accurate target identification, thereby fortifying defenses against potential adversarial threats. Techniques for secure transmission utilizing multiple satellites leverage coordinated communication across a constellation of satellites, thereby augmenting resilience against jamming and interception attempts. Concurrently, security-oriented RF methodologies concentrate on the protection of the RF spectrum through the application of advanced signal processing techniques, interference mitigation strategies, and the design of secure waveforms.

Safeguarding Satellite Communications: Issues, Challenges, and Solutions, First Edition.
Jianping An, et al.

11.1 ISAC-aided Secure Space Communication Techniques

11.1.1 Concept of ISAC

The swift expansion of interconnected devices and satellites has resulted in heightened congestion within the frequency spectrum allocated for wireless communications, thereby prompting network architects to explore opportunities for spectrum reuse to optimize the utilization of bands that were initially designated for alternative technologies. In recent years, the parallels between communication and sensing—evident in shared hardware components, antenna designs, and signal processing architectures—have catalyzed the emergence of ISAC.

ISAC denotes a sophisticated technological framework that integrates the traditionally distinct roles of sensing (for instance, radar, cameras, light detection and ranging [LiDAR]) and communication into a cohesive structure, thereby facilitating uninterrupted interaction between data acquisition and communication mechanisms. Through the shared utilization of spectrum, antennas, and signals, ISAC promotes the coordinated optimization of sensing and communication, employing a consolidated signal processing architecture for joint enhancement, and capitalizing on adaptive control methodologies to dynamically modify the priority assigned to sensing and communication contingent upon application requirements, thereby improving spectrum efficiency, system performance, and environmental responsiveness [1, 2].

11.1.2 Advantages and Challenges

11.1.2.1 Advantages

The benefits provided by ISAC for SCSs encompass mitigating the scarcity of RF resources, lowering the comprehensive expenses of the system, minimizing energy consumption, and facilitating the miniaturization of devices. A significant obstacle in wireless communication pertains to the restricted availability of the spectrum. ISAC contributes to the resolution of this issue by enabling the simultaneous utilization of the same spectrum for both sensing and communication, thereby augmenting overall spectrum efficiency [3]. Specifically, the robust directivity, minimal side-lobe levels, and anti-jamming features inherent in radar technology are instrumental in bolstering information security, enhancing transmission reliability, improving communication quality, and extending the coverage of LEO SCSs [4]. Furthermore, the sensing component of ISAC can partially elucidate the propagation environment, which has the potential to enhance the accuracy of CSI estimation and diminish the overhead associated with channel estimation. The acquired movement and locational data of high-mobility entities also aid LEO SCSs in refining their beam alignment methodologies and routing algorithms [5].

In the context of SCSs, ISAC can yield significant environmental insights, such as meteorological conditions or satellite positioning, which may facilitate the optimization of communication link performance and elevate service quality [6].

Conversely, the data acquired through communications significantly contributes to the enhancement of precise localization, real-time monitoring, and high-fidelity imaging, as well as the recognition of activities [5], which holds substantial promise in fortifying the security of LEO SCSs. For example, precise localization and monitoring establish robust underpinnings for operations related to debris elimination. Furthermore, high-fidelity imaging, which is augmented by AI, enables LEO SCSs to be optimized to considerably mitigate the likelihood of space collisions. The LEO SCSs are also anticipated to exhibit elevated classification accuracy, facilitating the detection of anomalies and imminent malfunctions through continuous environmental sensing.

11.1.2.2 Challenges

The implementation of ISAC introduces a series of supplementary challenges. The constraints encountered by ISAC in various wireless networks are similarly prevalent in SCSs. For example, it must explore comprehensive performance metrics that assess the trade-offs between communication and sensing capabilities, elucidate the fundamental limitations of ISAC under pragmatic conditions, and devise low-complexity yet precise signal processing algorithms for ISAC [2]. Furthermore, the dynamic mobility and technological constraints of satellites contribute to additional difficulties, such as limited synchronization precision, constrained channel coherence duration, high-velocity moving targets, and inadequate computational resources. ISAC systems are required to reconcile the competing demands of real-time data processing, communication fidelity, and environmental sensing. Achieving this will necessitate the development of advanced algorithms and network protocols capable of managing the inherently dynamic characteristics of such systems [7].

Moreover, the substantial quantities of data produced by ISAC, which encompass sensitive information, present additional obstacles in the safeguarding of SCSs. To protect the sensitive data generated by ISAC, the implementation of robust encryption and authentication protocols is imperative. Nevertheless, the real-time encryption of extensive data volumes can impose a significant computational strain on satellites. Furthermore, as the proliferation of satellite constellations leads to an increase in the number of satellites, the intricacy of managing secure communication escalates. The necessity for secure inter-satellite communication, data encryption, and seamless authentication among numerous satellites and ground stations necessitates the development of scalable security frameworks.

11.1.3 Key Applications of ISAC

With ISAC, satellites transcend mere communication functions to engage in environmental monitoring and sensing, as illustrated in Figure 11.1. In sectors such as military operations, maritime activities, meteorological surveillance, smart home technologies, and vehicular networks, ISAC employs LEO satellites to facilitate efficient data transmission and accurate sensing capabilities. For instance, within maritime and meteorological contexts, ISAC satellites furnish low-latency communication while monitoring oceanic and atmospheric conditions, thereby augmenting safety measures and disaster response strategies. In the realms of smart homes and vehicular networks, ISAC enhances IoT connectivity in remote locales and streamlines data transmission processes. In contrast to conventional SCSs, ISAC delivers superior spectrum sharing, interference mitigation, and optimization of data transmission through the integrated optimization of spectrum and hardware resources, thereby addressing the future demands of space-terrestrial integrated networks which require elevated throughput and minimal latency.

In the domain of security, ISAC markedly improves the resilience of SCSs against jamming and eavesdropping through advanced signal processing techniques and refined transmission methodologies. ISAC diminishes the vulnerability to unauthorized interception by employing a combination of sophisticated BF, artificial

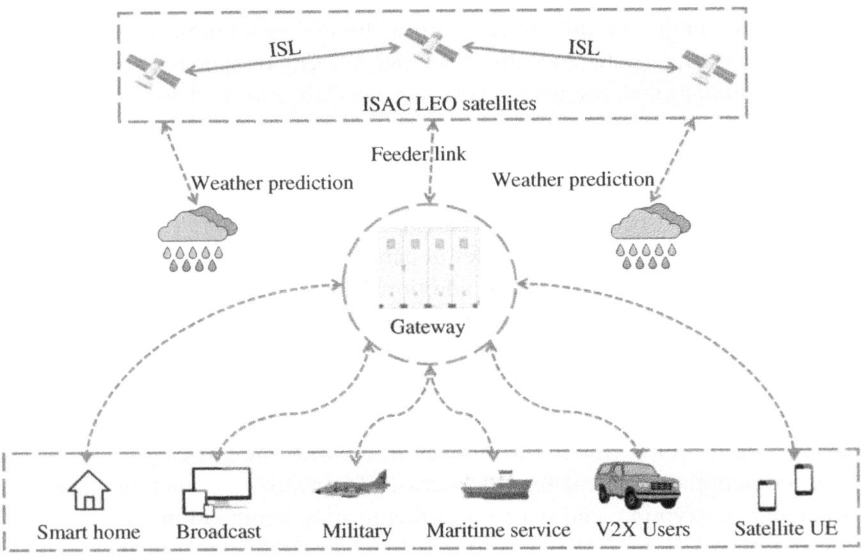

Figure 11.1 Application scenarios of ISAC in satellite communications.

jamming mechanisms, and secure precoding, while dynamically refining transmission protocols following the CSI of potential eavesdroppers [8]. Furthermore, through the integration of sensing data, ISAC empowers satellites to accurately identify imminent threats and optimize the orientation of their beams, thereby safeguarding the confidentiality and jamming resistance of tactical communication channels. In addition, ISAC's capability to monitor and respond to fluctuating network conditions facilitates rapid recovery of satellites from disruptions or malicious attacks, thereby assuring persistent reliability and security of communication links, particularly in high-stakes scenarios where both security and operational continuity are of utmost importance.

11.2 CV-aided Secure Space Communication Techniques

11.2.1 Description of CV

Recently, CV-aided communication schemes have been advanced for mmWave or THz transmission systems, wherein LoS propagation is of paramount importance [9, 10]. In contrast to ISAC, CV-assisted methodologies predominantly concentrate on the extraction, recognition, and estimation of critical information regarding the static topology of the system, which includes the positions of terminals, their inter-distances, and their quantities. Furthermore, the continuous monitoring of their velocity, directional vectors, and movement trajectories serves to further enhance system optimization. The primary objective of these methodologies is to refine the design and optimization of wireless systems, thereby facilitating improved resource scheduling, allocation, and the development of sophisticated algorithms and protocols.

CV is integral to the design of contemporary communication systems, integrating AI, ML, DL, signal processing, and optical engineering. It facilitates the extraction and examination of visual data to emulate human perception, thereby enhancing both the efficiency and security of communication processes. The methodology generally encompasses several critical phases: data acquisition, which involves the utilization of diverse sensors, such as conventional cameras, infrared cameras, depth sensors, radar, and hyperspectral sensors, to amass image or video data; data preprocessing, aimed at improving image quality through noise reduction, contrast adjustment, and achieving uniformity for enhanced analytical capability; feature extraction, wherein pivotal visual information is derived employing techniques such as convolutional neural networks (CNNs) or Vision Transformers (ViTs) to pinpoint key features, edges, and patterns; and pattern recognition, which classifies and identifies objects through ML or DL methodologies. In the realm of wireless communication, these methodologies are employed for applications including obstacle detection, BF, and channel state analysis, thereby augmenting system performance in intricate environments such as mmWave and THz communication.

11.2.2 Feasible Techniques

This particular subsection will examine the viable methodologies pertaining to the CV techniques that can be employed via satellite systems: optical imaging apparatus, CV processing facilities, and various CV methodologies.

11.2.2.1 Optical Camera

To enhance the efficacy of CV-based space target monitoring systems, the integration of extensive coverage and precise close-range monitoring capabilities, facilitated by advancements in microelectronics technology, has rendered small satellites increasingly pivotal in the domain of space target surveillance. These satellites are equipped with a variety of sophisticated instruments designed for the detection of celestial targets, including high-resolution imaging systems, infrared sensing technology, hyperspectral analysis tools, radar systems, and LiDAR apparatus.

Although high-resolution imaging devices are capable of delivering precise visual representations for target identification, the processes of sensing, transmission, and data processing are inherently time-consuming. Infrared technology encompasses a wider array of applications. While it exhibits a reduced vulnerability to interference and deceptive practices, it remains susceptible to the effects of sunlight and necessitates sophisticated algorithms for effective data processing. Hyperspectral instruments possess the ability to ascertain the chemical composition and structural attributes of targets; however, they are characterized by suboptimal image quality. Future advancements in space target monitoring technologies are anticipated to evolve and diversify through the enhancement and modernization of various detection apparatuses.

11.2.2.2 CV Processing Center

To augment the computational capabilities of the satellite system and render it versatile across diverse scenarios, the CV processing center has progressively adopted sophisticated hardware platforms to satisfy the escalating computational requirements of satellite applications.

The Nvidia Jetson serves as a robust, low-power AI module of high performance and compact dimensions, incorporating a powerful Nvidia Pascal graphic processing unit (GPU) specifically designed for satellite applications. It exhibits exceptional reliability alongside resistance to radiation, rendering it suitable for satellite image enhancement, object detection, and tracking purposes. As a programmable logic device, FPGAs provide enhanced adaptability and parallel computational capabilities. These devices can be employed in satellite systems utilizing AI algorithms, exemplified by Phi-sat-1 and SkyLabs PicoSkyFT. This reconfigurable processing platform has been implemented as an onboard computer or payload processor for nanosatellites, accommodating a variety of communication interfaces and general-purpose input/output functionalities. Multi-core parallel processors are adept at facilitating parallel computing and multitasking operations. Their functionality is characterized by high efficiency and throughput achieved through the collaborative engagement of multiple central processing unit (CPU) cores. While they serve as pivotal computing platforms for hub nodes, their implementation necessitates intricate software design and debugging processes.

11.2.2.3 Techniques of CV

Traditional space communication systems, which are predicated on RF signals, encounter substantial constraints concerning the transmission of extensive data volumes over considerable distances while simultaneously avoiding interference or signal attenuation. To mitigate the shortcomings associated with RF technologies, CV-based communication systems emerge as a highly promising alternative with the capacity to fundamentally transform the domain of space communications. By harnessing sophisticated image processing and AI methodologies, including ML algorithms, CV can proficiently extract pertinent information from visual data and convey it to terrestrial stations or spacecraft. This innovative technology has already exerted a profound influence on a variety of space missions, encompassing Earth observation, remote sensing initiatives, and explorations of deep space.

CV-based methodologies for satellite detection and tracking are increasingly receiving scholarly attention, particularly within the domains of space exploration and satellite communications. Established techniques, encompassing object detection, classification, and real-time trajectory forecasting, are employed to precisely and effectively identify and monitor satellites. For example, algorithms such as the you only look once (YOLO) methodology, faster region-based CNN (R-CNN), and detection transformers are extensively utilized owing to their rapidity and precision in identifying small or irregularly shaped satellites [11]. Furthermore, classification models, notably CNNs, have been devised to differentiate satellites from other entities.

11.2.3 Applications of CV in Satellite Communication

Satellite communications are progressively integrating THz and FSO technologies to satisfy the escalating demand for high-speed data transmission within advanced broadband Internet mega-constellations. Nonetheless, these high-frequency technologies necessitate the utilization of highly directional beams, presenting a significant challenge in the dynamic milieu of outer space. CV-based joint sensing and communication (JSC) emerges as a promising solution to enhance space communications. However, conventional JSC methodologies are inadequately equipped for the unique conditions of space. Passive radar systems exhibit inefficacy in detecting signal fluctuations in the outer space environment, while active radar introduces substantial delays and engenders security concerns. Closed-loop JSC architectures depend on feedback mechanisms, resulting in elevated latency attributable to the considerable distances and relative velocities among satellites. Furthermore, the proliferation of satellites amplifies the susceptibility of radar signals to eavesdropping. To mitigate these challenges, an open-loop joint space sensing and communication (JSSC) framework predicated on CV is proposed, with its fundamental architecture illustrated in Figure 11.2 [12].

The prominent characteristics of CV-aided methodologies, which prove advantageous for enhancing the security of LEO SCSs, can be delineated as follows. A fundamental attribute of CV in satellite communications is its unidirectional sensing capability. The implementation of CV technology empowers satellites outfitted with optical imaging devices or multispectral sensors to conduct real-time surveillance of terrestrial, aerial, or extraterrestrial targets [12]. Through the application of optical imaging equipment and sophisticated image processing methodologies, satellites are capable of passively accumulating environmental and situational intelligence without necessitating RF transmissions. This strategy

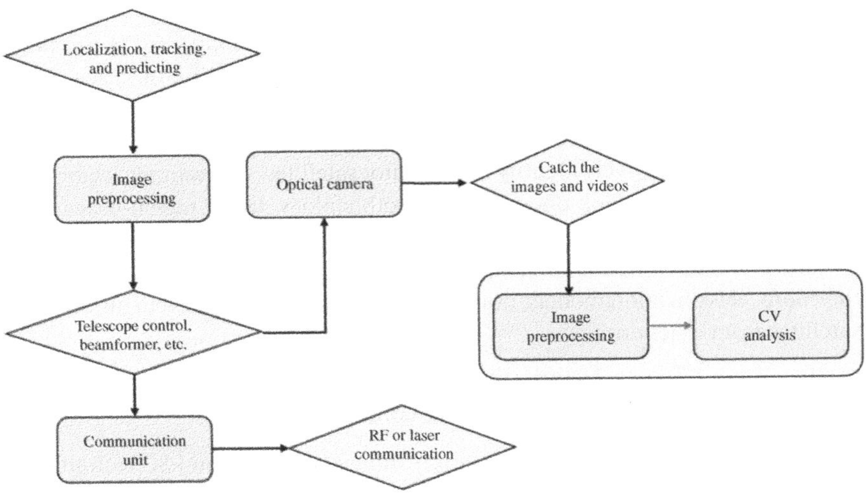

Figure 11.2 An open-loop integrated spatial sensing and communication system model based on CV.

not only mitigates the likelihood of detection by hostile entities but also diminishes energy consumption and bandwidth demands, consequently enhancing the overall efficacy and clandestineness of communication operations [13].

Another pivotal application of CV in the realm of satellite communication pertains to the detection of hostile targets and non-partner entities. Through the utilization of advanced optical processing algorithms, it becomes feasible to detect, identify, and differentiate between authorized and unauthorized subjects within their observable environment. Consequently, secure and covert information transmission methodologies can be devised and executed to mitigate the risks of potential eavesdropping and disruptions [14]. Table 11.1 delineates a comparative analysis of three methodologies with respect to their levels of covertness, power consumption, and operational frameworks. In contrast to radar and RF communications that operate within the RF spectrum, the open-loop CV-based JSSC concept exhibits promise for applications in outer space. The CV-oriented communication framework can adeptly extract crucial information from imagery for transmission to spacecraft by employing sophisticated image processing and AI methodologies [12].

11.2.4 Future Development Directions of CV

The prospective advancement of CV within satellite communication frameworks is poised to significantly augment multiple dimensions of extraterrestrial operations. A particularly auspicious avenue is SSA, wherein CV technologies scrutinize optical or radar imagery to identify possible hazards, including non-cooperative satellites or orbital debris. This capability can furnish preemptive alert data to avert collisions or disruptions.

Another paramount domain pertains to anti-deception methodologies. CV can facilitate the detection of camouflage patterns employed by adversarial satellites, thereby enhancing the system's resilience against deceptive tactics and guaranteeing secure communication. This capability will be essential for preserving the integrity of satellite communication networks within increasingly intricate extraterrestrial environments. Furthermore, the amalgamation of RL with

Table 11.1 Comparison of the open-loop CV-based JSC model with RF and radar-based object detection.

Approaches	Working methodology	Power consumption	Latency	Covertness
Radar	Closedloop	High	High	Poor
RF communication	Closedloop	High	High	Poor
CV-based concept	Openloop	Low	Low	High

CV technology presents substantial promise. Through the integration of CV with RL algorithms, satellites may dynamically modify their communication strategies informed by historical data, thereby augmenting efficiency in fluid and unpredictable contexts.

11.3 Multi-satellite Secure Transmission Techniques

11.3.1 Multi-satellite System Characteristics

In contemporary satellite communication frameworks, the hybrid architecture comprising LEO, MEO, and GEO has emerged as a significant paradigm. LEO satellites facilitate communication characterized by minimal latency and elevated data transmission rates, owing to their proximity to the Earth's surface. MEO satellites achieve an optimal equilibrium between coverage area and latency, whereas GEO satellites ensure extensive, uninterrupted coverage across vast regions. The ISL topology inherent in this hybrid architecture interconnects these satellites, thereby facilitating data transfer across various orbital altitudes, as illustrated in Figure 11.3.

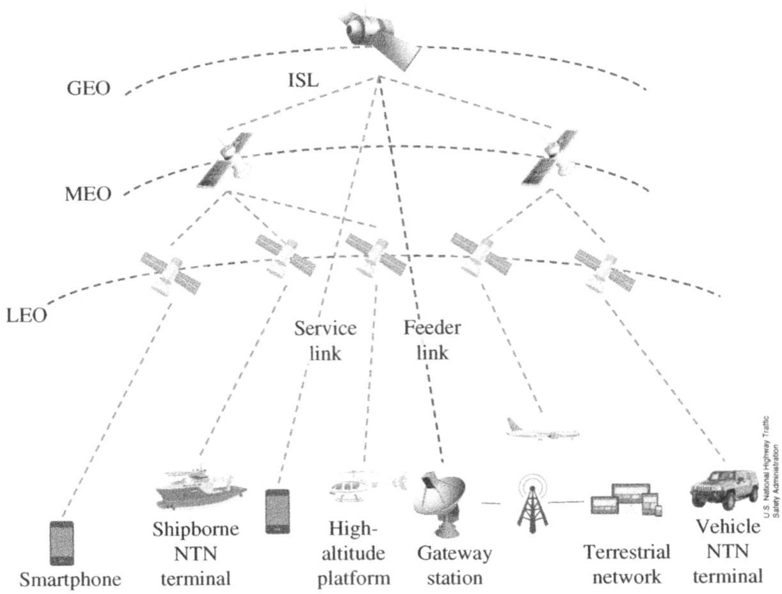

Figure 11.3 Multi-satellite collaborative communication scenario.

The inherent dynamism and elevated mobility associated with multi-satellite systems present considerable challenges to security paradigms. Link switching frequently transpires as satellites traverse their respective orbits, a phenomenon that can be exploited by nefarious entities who may endeavor to disrupt the handover process or illicitly gain access during the transitional phase. Furthermore, delay jitter, which is an additional repercussion of satellite mobility, may also undermine security measures. Fluctuating delays have the potential to interfere with the synchronization of security protocols, particularly those employed for authentication and encryption purposes. For example, in a system wherein time-sensitive cryptographic procedures are contingent upon precise time synchronization, the presence of delay jitter may precipitate failures in these operations, thereby rendering the communication susceptible to various forms of attacks.

11.3.2 Multi-satellite Secure Transmission Technology

11.3.2.1 ISL Security Enhancement Technology

The inherent dynamism and elevated mobility associated with multi-satellite systems present considerable challenges to security paradigms. Link switching frequently transpires as satellites traverse their respective orbits, a phenomenon that can be exploited by nefarious entities who may endeavor to disrupt the handover process or illicitly gain access during the transitional phase. Furthermore, delay jitter, which is an additional repercussion of satellite mobility, may also undermine security measures. Fluctuating delays have the potential to interfere with the synchronization of security protocols, particularly those employed for authentication and encryption purposes. For example, in a system wherein time-sensitive cryptographic procedures are contingent upon precise time synchronization, the presence of delay jitter may precipitate failures in these operations, thereby rendering the communication susceptible to various forms of attacks.

11.3.2.2 Multi-satellite Collaborative Secure Transmission Mechanism

Distributed key management, exemplified by blockchain-based swarm key negotiation, constitutes a fundamental component of secure transmission in multi-satellite collaborative systems. Within a multi-satellite framework, the reliance on a centralized key management system poses a significant risk as it represents a potential single point of failure. The application of blockchain technology facilitates the decentralized negotiation of keys among satellites, as illustrated in Figure 11.4. Each satellite within the swarm is empowered to participate in the key negotiation process, while the blockchain framework guarantees the integrity and security of the key-sharing mechanism. Furthermore, the secure transmission of redundant pathways represents another critical mechanism. The implementation of multipath piecewise encryption alongside cross-satellite authentication

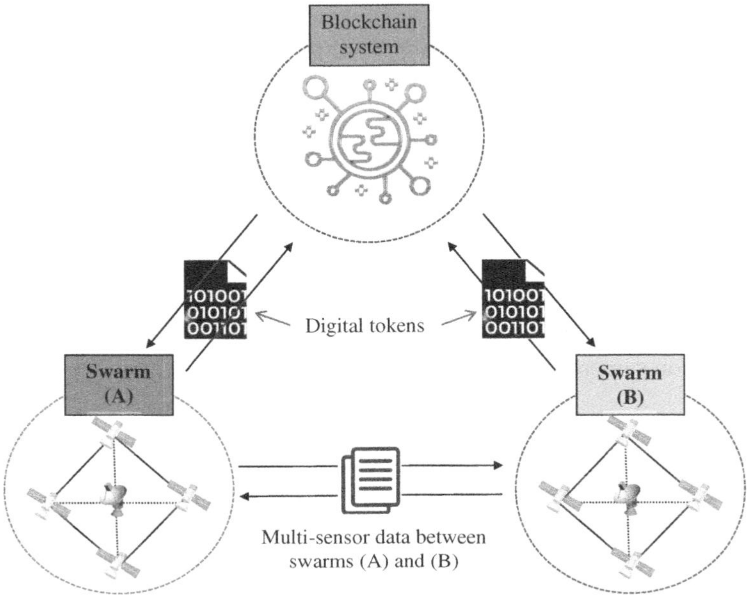

Figure 11.4 Satellite swarm connection management and protection based on blockchain.

entails segmenting the data into multiple components and encrypting each segment independently. Subsequently, each segment is transmitted via diverse pathways among the satellites, and cross-satellite authentication is employed to ensure the integrity of the data throughout the transmission process. Additionally, cooperative anti-interception technology, exemplified by multi-star joint BF interference avoidance, is of paramount importance. By synchronizing the BF capabilities of multiple satellites, the system can effectively direct the communication signal toward the designated receiver while circumventing regions where interception is probable. This strategic approach significantly complicates the efforts of unauthorized entities to intercept the communication.

11.3.3 Key Applications of Multi-satellite Secure Transmission Technology

Multi-satellite secure transmission technology facilitates the creation of a communication channel that is both exceptionally secure and dependable. In the context of military operations conducted in isolated regions where conventional communication frameworks might be lacking or susceptible to threats, the LEO/MEO/GEO hybrid architecture guarantees that military personnel are able to engage in secure communication. Advanced link authentication and anti-jamming methodologies serve to safeguard communication against the risks of interception and disruption by adversarial forces.

Furthermore, the extensive utilization of remote IoT necessitates a substantial interconnectivity of devices, thereby propelling the evolution of cooperative multi-satellite technologies. By capitalizing on the spatial diversity attributes inherent in multiple satellites, the system can markedly improve its resistance to jamming, thereby effectively alleviating the repercussions of the inherent openness of satellite-to-ground communication channels and the predictability associated with satellite trajectories. Consequently, this fortifies defenses against disruptive broadband malicious jamming emanating from external entities [15].

11.4 Security-oriented RF Techniques

11.4.1 Security-oriented RF Technology Background

RF technology constitutes the foundational element of the physical layer in SCSs, while threats such as eavesdropping, jamming, and camouflage emerge as significant concerns. Conventional encryption methodologies, although integral to the protection of communications, present their own array of challenges. The management of cryptographic keys, for example, represents a multifaceted and resource-demanding endeavor. Guaranteeing the secure generation, distribution, and preservation of cryptographic keys proves to be a formidable task, particularly within extensive and dynamic communication infrastructures. Moreover, the computational burden imposed by traditional encryption frameworks can be considerable, particularly for devices constrained by limited processing capabilities and energy supplies. The principal objective of security-focused RF technology is to equip communication systems with the capacity to withstand jamming attempts, thwart eavesdropping activities, and identify forgery originating from the signal source.

11.4.2 Security-oriented RF Techniques

One pivotal methodology for augmenting PLS is the implementation of AN, as illustrated in Figure 11.5. With the assumption of possessing prior knowledge regarding the CSI of the eavesdropper, AN can be meticulously engineered utilizing BF techniques to attenuate the quality of the eavesdropper's channel, consequently bolstering the security of legitimate communications. In the context of a multiple-antenna communication system, AN can be architecturally structured to be orthogonal to the signal designated for the legitimate recipient within the spatial domain, thereby guaranteeing that authorized communication remains

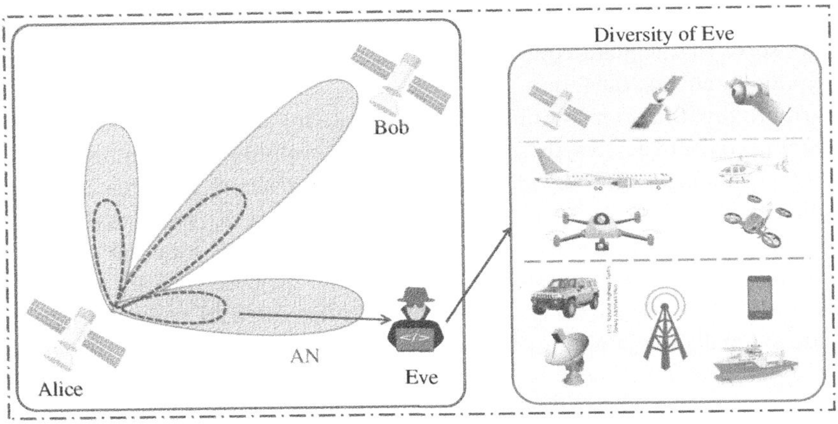

Figure 11.5 Secure BF based on AN.

unimpeded while effectively obstructing the eavesdropper's reception. The optimization of AN-driven secrecy schemes is predominantly centered on ascertaining the optimal power distribution for AN to maximize the secrecy capacity of the legitimate transmission. Nonetheless, owing to the necessity for prior knowledge of the eavesdropper's channel, scholars have been actively exploring methodologies to fortify information transmission security via AN injection without necessitating the acquisition of CSI.

Another pivotal method for enhancing PLS is channel fingerprint authentication, which capitalizes on the distinctive attributes of CSI as a dynamic identifier for the purpose of authentication. The communication channel that exists between the transmitter and the receiver possesses inherent characteristics that may vary over time, influenced by factors such as environmental conditions, the mobility of the communicating devices, and external interference. Through the continuous observation and examination of the CSI, it becomes feasible to generate a unique fingerprint representative of the communication channel. This distinctive fingerprint may be employed to verify the authenticity of the communication link. Should an unauthorized device attempt to intercept or replicate the communication, the CSI-based fingerprint will diverge from the anticipated pattern, thereby alerting the legitimate parties to the possible security compromise.

Beyond the domain of PLS, RF identification (RFID) methodologies furnish an ancillary layer of safeguarding by capitalizing on the innate hardware anomalies present in transmitting apparatus as distinctive identifiers. The certification of device uniqueness predicated on the hardware imperfections of transmitters constitutes a pivotal element of RF-centric security paradigms. The hardware of the receiver is inherently flawed and manifests certain intrinsic defects, including phase noise and I/Q imbalance. Such deficiencies are unique to each device and can serve as a variant of a "fingerprint" for the purpose of device identification. For instance, phase noise, characterized by the stochastic variations in the phase of a transmitted signal, diverges from one transmitter to another. Through meticulous measurement and analysis of these hardware-specific anomalies, it becomes feasible to achieve a distinct identification of a transmitter and to ascertain that only authorized devices are engaged in the communication process.

11.4.3 Key Applications of Security-oriented RF Techniques

DL has emerged as a formidable instrument to enhance RF security. In the realm of detecting anomalous interference, models predicated on DL can be systematically trained to identify patterns in the RF signals that are received. These models possess the capability to assimilate the normative behavior of the RF environment, encompassing the attributes of both legitimate signals and ambient noise. Upon the occurrence of anomalous interference, such as the introduction of a novel jamming signal, the DL model is adept at promptly recognizing deviations from the established patterns. Following the detection, the system is able to implement adaptive countermeasures. For instance, it may recalibrate the transmission parameters, including the frequency band, power level, or modulation scheme, to alleviate the adverse effects of interference and preserve the integrity of the communication.

THz communications present the potential for rapid data transmission, yet they simultaneously introduce novel security concerns. Investigating secure RF architectures within this frequency domain is of paramount importance. Considering the distinctive propagation characteristics of THz waves, which include significant atmospheric attenuation and heightened sensitivity to environmental influences, it may be necessary to modify conventional RF security methodologies. Innovative security-focused RF architectures in THz communications could entail the creation of advanced PLS strategies specifically designed for the THz milieu, in addition to capitalizing on the inherent properties of THz waves to bolster security, such as employing the short-range and highly directional features of THz signals to restrict the extent of possible eavesdropping.

References

1 Fan Liu, Christos Masouros, Athina P Petropulu, et al. Joint radar and communication design: Applications, state-of-the-art, and the road ahead. *IEEE Trans. Commun.*, 68(6):3834–3862, Jun. 2020.

2 An Liu, Zhe Huang, Min Li, et al. A survey on fundamental limits of integrated sensing and communication. *IEEE Commun. Surv. Tutor.*, 24(2):994–1034, Second Quart. 2022.
3 Zesong Fei, Xinyi Wang, Nan Wu, Jingxuan Huang, and J, Andrew Zhang. Air-ground integrated sensing and communications: Opportunities and challenges. *IEEE Commun. Mag.*, 61(5):55–61, 2023.
4 Zhiyong Feng, Zixi Fang, Zhiqing Wei, et al. Joint radar and communication: A survey. *China Commun.*, 17(1):1–27, Jan. 2020.
5 Danny Kai Pin Tan, Jia He, Yanchun Li, et al. Integrated sensing and communication in 6G: Motivations, use cases, requirements, challenges and future directions. In *2021 1st IEEE International Online Symposium on Joint Communications & Sensing (JC&S)*, pages 1–6, Dresden, Germany, 23–24 Feb. 2021. IEEE.
6 Zhiqing Wei, Hanyang Qu, Wangjun Jiang, Kaifeng Han, Huici Wu, and Zhiyong Feng. Iterative signal processing for integrated sensing and communication systems. *IEEE Trans. Green Commun. Netw.*, 7(1):401–412, 2023.
7 Yifeng Xiong, Fan Liu, Yuanhao Cui, Weijie Yuan, Tony Xiao Han, and Giuseppe Caire. On the fundamental tradeoff of integrated sensing and communications under gaussian channels. *IEEE Trans. Inf. Theory*, 69(9):5723–5751, 2023.
8 Xue Zhang, Ruibo Wang, Bodong Shang, and Mohamed-Slim Alouini. Joint robust secure beamforming designs for ISAC-enabled LEO satellite systems. In *ICC 2024—IEEE International Conference on Communications*, pages 1182–1188, 2024.
9 Yu Tian, Gaofeng Pan, and Mohamed-Slim Alouini. Applying deep-learning-based computer vision to wireless communications: Methodologies, opportunities, and challenges. *IEEE Open J. Commun. Soc.*, Dec. 2020.
10 Nuria Gonzalez-Prelcic, Anum Ali, Vutha Va, and Robert W Heath. Millimeter-wave communication with out-of-band information. *IEEE Commun. Mag.*, 55(12):140–146, Dec. 2017.
11 Joseph Redmon, Santosh Divvala, Ross Girshick, and Ali Farhadi. You only look once: Unified, real-time object detection. In *Proceedings of the IEEE Conference on Computer Vision and Pattern Recognition (CVPR)*, pages 779–788, 2016.
12 Han Yu, Zizheng Hua, Xiaqing Miao, et al. Computer vision-based joint space sensing and communication systems: Non-source, autonomy, and low latency. *IEEE Wirel. Commun.*, 30(6):60–68, 2023.
13 Raúl Mur-Artal and Juan D Tardós. ORB-SLAM2: An open-source SLAM system for monocular, stereo, and RGB-D cameras. *IEEE Trans. Robot.*, 33(5):1255–1262, Oct. 2017.
14 Alexey Bochkovskiy, Chien-Yao Wang, and Hong-Yuan Mark Liao. YOLOv4: Optimal speed and accuracy of object detection, arXiv preprint arXiv:2004.10934, 2020.
15 Shuqi Qiao, Yizhe Shao, Guodong Kang, et al. Code-assisted broadband jamming suppression in multi-satellite collaborative secure communication. *IEEE Trans. Aerosp. Electron. Syst.*, 61, pages 1–14, 2024.

12

Secure-oriented Mega-constellation

The designation "Secure Oriented Mega-constellation" pertains to a space-based framework that is constituted of numerous satellites, primarily aimed at the enhancement of security. An exemplary instance of such an endeavor is the "Secure Connectivity System" proposed by the European Commission, which is conceptualized as a multi-orbit constellation intended to facilitate satellite-based Internet services within vulnerable regions across Europe and Africa. Furthermore, the system is planned to deliver quantum-encrypted communications to governmental entities, consequently bolstering the resilience of critical infrastructure and fostering the secure transmission of information. Such extensive constellations typically employ an all-link security architecture, which ensures the integrity of data transmission and the dependability of the system. These systems can subsequently be utilized to confront the complex security challenges that may arise in military, cyber, and space sectors. Technological strategies may be employed to mitigate these threats, such as augmenting cybersecurity measures and enhancing situational awareness capabilities. Conversely, the threats may also be tackled through management strategies, encompassing the refinement of policies and regulations as well as the fortification of international collaboration.

12.1 Mega-constellation Design

Mega-constellations represent extensive networks of satellites meticulously engineered to operate in unison, thereby ensuring comprehensive global communication coverage [1]. These constellations typically comprise hundreds or thousands of small satellites situated in LEO, strategically arranged to deliver near-continuous coverage of the Earth's surface. By leveraging this intricate constellation of satellites, organizations are able to furnish high-speed Internet

Safeguarding Satellite Communications: Issues, Challenges, and Solutions, First Edition.
Jianping An, et al.

and ancillary services in areas that have historically been marginalized, such as isolated regions and oceanic expanses, where terrestrial infrastructure is either nonexistent or economically impractical.

12.1.1 Current Status and Design Features

1. **Current Research Status in Mega-constellations:** Globally considered, international entities have initiated an early and consistent engagement in the advancement of extensive satellite constellations. SpaceX's Starlink represents the most notable instance of this trend, with aspirations to deploy 41,927 satellites aimed at delivering broadband Internet services to users globally. The achievement of Starlink in realizing this ambitious objective can be ascribed to the interplay of two pivotal factors: The impact of Elon Musk, a distinguished individual within the global technology landscape, and the technological capabilities of SpaceX. Consequently, there has been a significant augmentation in the quantity of satellites in orbit, with services now accessible in various regions worldwide, thereby markedly enhancing network coverage in isolated locales. In a comparable manner, Amazon has articulated plans to implement a comprehensive constellation of 3,236 satellites, a strategic initiative that effectively positions the corporation within the swiftly evolving broadband Internet service sector. Likewise, OneWeb has focused its ambitions on attaining global influence through the rollout of 1,980 satellites, thereby firmly entrenching its presence in the domain of global satellite communications. Both Amazon and OneWeb possess substantial financial resources and technological expertise, thereby positioning themselves as formidable competitors in the quest for supremacy in the global satellite Internet arena.

   In the context of national initiatives, there exists a remarkable dynamism within the sphere of large-scale satellite constellations. An illustrative example is the "Galaxy" initiative commenced by Galaxy Space, which is set to deploy 650 satellites aimed at facilitating broadband Internet connectivity. This endeavor exemplifies a wider phenomenon, as Geely, another prominent entity within this industry, is engaging in a comparable project that remains unnamed but is anticipated to launch 168 satellites, similarly aimed at delivering broadband Internet services. This highlights the ongoing trend of transnational integration and advancement. The Xingyun initiative, managed by the China Aerospace Science and Technology Corporation, emphasizes narrowband IoT technology and is scheduled to deploy 80 satellites, which are expected to serve a crucial function in IoT applications. Conversely, the Hongyun initiative focuses on broadband Internet provision and has a projected launch of 156 satellites. The Hongyan initiative, overseen by the

Aerospace Science and Technology Group, is committed to broadband Internet services and is planning the deployment of 300 satellites.

According to the data presented in Table 12.1, the predominant objective of the majority of contemporary large-scale satellite constellation initiatives is the provision of broadband Internet service. This observation underscores the pressing global demand for high-speed and reliable Internet connectivity, particularly in isolated regions and in environments such as maritime and aerial domains that are challenging to reach through conventional terrestrial networks. Concurrently, the quantity of satellites proposed by various companies exhibits significant variability, which correlates with the strategic foresight, financial capabilities, and technological assets of each organization, thereby highlighting the heterogeneity of market competition.

The emergence of extensive constellations signifies considerable progress in the realm of space technology, carrying substantial ramifications for the international communications framework. As these constellations are progressively designed and put into operation, they are likely to offer improved

Table 12.1 The evolution of space-based laser communications.

Characteristics	Single-layer constellation	Multilayer constellation
Design features	All satellites are positioned on the same orbit with a high inclination angle (e.g., LEO, 300–2,000 km)	Satellites are distributed across multiple orbital layers (LEO + MEO/GEO or LEO with non-inclined orbits)
Advantages	- Simple structure - Lower cost - Short transmission delay (approximately 20–50 ms)	- Higher redundancy, increased anti-jamming ability - Optimized orbital configuration (low orbit for low latency, high orbit for broad coverage) - Divided service levels, with support for extensive traffic and heavier payloads
Limitations	- Limited satellite coverage due to orbital constraints - Requires more satellites to meet user demands - Single-layer orbit increases failure risk	- System complexity, higher cost
Representative systems	Starlink (Phase 1), OneWeb Constellation	Enhanced version: Starlink V2.0 (supports ISLs) China's "Hongmeng" Constellation plan

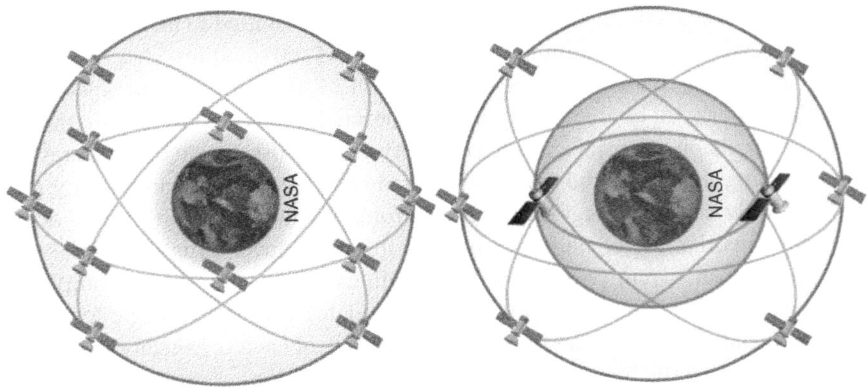

(a) (b)

Figure 12.1 Schematic of the basic architecture of the mega-satellite constellation. (a) Single-layer constellation and (b) multilayer constellation.

network services to users on a global scale, thereby facilitating the expansion of the IoT, intelligent transportation systems, and related fields. Nevertheless, this advancement is accompanied by a series of challenges, which include competition for orbital assets, the increase of space debris, and the necessity for comprehensive cybersecurity protocols to safeguard the safety and integrity of these systems.

2. **Design Features:** The foundational design principle underlying mega-constellations is to enhance performance and extend coverage via a distributed satellite network. As illustrated in Figure 12.1, this network configuration is categorized into two primary classifications: Single-layer constellations and multilayer constellations.

Single-layer Constellation: A single-layer mega-constellation constitutes a satellite architecture wherein a substantial quantity of satellites is positioned within the identical orbital structure. This configuration presents the advantage of relative simplicity in both design and operation, akin to athletes arranged on a uniform runway, facilitating efficient deployment and management. Nevertheless, this simplicity is accompanied by inherent constraints. The restricted allocation of orbital slots results in inadequate "substitutability" among satellites; when a satellite becomes nonoperational, it can generate a "gap" in coverage, thereby jeopardizing the stability of the entire system. To alleviate these prospective vulnerabilities, it is imperative to augment the number of satellites, which will inevitably lead to escalated launch and maintenance expenditures. Concurrently, signal transmission constrained by orbital parameters in certain specific locales, such as high-latitude polar regions or isolated mountainous terrains, is also susceptible to "jamming," thereby complicating the achievement of uninterrupted global coverage.

Multilayer Constellation: Multilayer mega-constellations comprise intricate systems that deploy satellites across multiple orbital strata, including LEO, MEO, and GEO. By systematically arranging these satellites across varying layers, a significantly more extensive coverage area is accomplished, thereby enhancing the scope of communication services. The satellites positioned in LEO facilitate communication characterized by high temporal efficiency and minimal latency, whereas those situated in the MEO and GEO layers contribute superior coverage and advanced communicative functionalities. The inherent redundancy of satellites within each orbital layer guarantees the continuity of service in the event of a failure of a satellite in an alternate layer, thus bolstering the overall stability of the system. However, the design of such a constellation entails rigorous technical specifications, encompassing precise orbital configurations, collaborative satellite control mechanisms, and complex launch and maintenance protocols, which necessitate a considerable initial financial commitment. Nevertheless, the prolonged operational viability of the system may yield substantial economic returns.

12.1.2 Key Considerations and Challenges

One of the principal design considerations for a mega-constellation pertains to the meticulous arrangement of satellites to ensure uninterrupted coverage and consistent communication links [2]. This necessitates the precise computation of orbital parameters, which encompass altitude, inclination, and phasing, to eliminate coverage gaps and mitigate the potential for collisions among satellites. These constellations frequently utilize ISLs to facilitate communication between satellites, thereby diminishing dependence on terrestrial ground stations and permitting data to be transmitted through the network of satellites themselves. Such an approach can markedly enhance the speed and reliability of communication, especially in remote regions or areas affected by disasters.

Another critical factor in the conceptualization of a mega-constellation pertains to the choice of the optimal satellite platform. The satellites are required to be compact, lightweight, and exceptionally efficient, equipped with robust onboard systems for communication, propulsion, and power regulation. This is essential for minimizing both launch and operational expenditures while guaranteeing that each satellite is capable of executing its designated functions, such as data transmission, orbital position maintenance, and collision avoidance.

The comprehensive design of the mega-constellation must exhibit a high degree of scalability [3]. As the demand for services escalates, the quantity of satellites within the constellation may be augmented, thereby facilitating the system's expansion to address emerging requirements. This inherent scalability constitutes one of the fundamental characteristics of mega-constellations, permitting their adaptation to the dynamic technological and commercial landscape.

The formulation of a secure mega-constellation constitutes a sophisticated and intricate endeavor, necessitating the amalgamation of cutting-edge technologies across diverse domains. These technologies not only facilitate effective communication and operation of the constellation but also emphasize the system's security, scalability, and resilience against an array of potential threats. The subsequent sections delineate several critical elements pertinent to the design of a mega-constellation, with a particular emphasis on security considerations:

1. **The Tracks Compete with the Spectrum Resources:** The deployment of extensive satellite constellations has resulted in a significant increase in congestion within the near-Earth orbital environment, thereby presenting a formidable challenge to the optimal utilization of orbital assets. The ITU, as the global authority governing spectrum allocation, faces an unparalleled level of exigency to devise appropriate frequency bands for a multitude of satellite constellations within the limited spectrum resources, aimed at mitigating the risk of mutual interference. Simultaneously, the surge in the number of satellites in orbit has contributed to a notable rise in the likelihood of satellite collisions, emphasizing the necessity for dynamic coordination among satellites to guarantee their safe operational conduct.
2. **Intersatellite Communication and Network Management:** In the domain of interstellar communications, laser interstellar links have emerged as a pivotal means of communication for extensive satellite constellations, exemplified by Starlink V2.0, due to their inherent advantages of elevated transmission speeds and minimal power requirements. Nevertheless, the dynamic movement of orbiting satellites results in a continual alteration of the constellation's topology, thereby rendering the optimization of routing for laser interstellar links a significant challenge. Current scholarly investigations emphasize the imperative of rapidly and accurately determining optimal routes to facilitate efficient data transmission amidst a dynamically varying topology. Within multilayer constellations, the intricacies of network management are heightened, attributable to the diverse characteristics and operational functions of satellites situated in various orbital strata, necessitating sophisticated traffic scheduling across multiple protocol layers.
3. **Sustainability Issues:** Sustainability constitutes a paramount concern in the advancement of extensive satellite arrays, a matter that cannot be overlooked. The incremental addition of spacecraft invariably leads to a heightened incidence of malfunctioning spacecraft and an accumulation of space debris, which poses a threat to operational spacecraft, thereby perpetuating a deleterious cycle of debris proliferation. Consequently, it is imperative to refine the design and control strategies to minimize energy consumption and prolong the operational lifespan of the propulsion system, all while satisfying the propulsion demands of the spacecraft, thereby ensuring the sustained and stable functionality of the spacecraft over the long term.

12.2 Secure Cross-orbit Space Communication Methods

In consideration of the forthcoming proposals for mega-constellation launches, it is anticipated that the quantity of operational satellites in orbit will escalate to approximately 50,000 within a decade, resulting in an unparalleled magnitude of LEO SCSs. Consequently, the operators of these mega-constellations experience a substantial computational challenge, as they are required to oversee and administer the operational conditions and various functionalities of hundreds or even thousands of satellites in real time. A minor computational or command error could significantly disrupt this extensive network. Moreover, extensive LEO SCSs necessitate a multitude of ground stations and gateways. The authors of [4] project that roughly 123 ground station locations and 3,500 gateway antennas will be essential for the 4,400-satellite iteration of Starlink to near the throughput capabilities. Such a large-scale implementation will demand highly automated and secure management systems.

Furthermore, the exchange of information between mega-constellations and terrestrial stations is contingent upon the inter-satellite network established through ISLs within the constellation. This paradigm contrasts with that of conventional small-scale constellations, wherein information exchange is facilitated via satellite-ground links or through the relay capabilities of GEO satellites. Although the implementation of the ISL framework has resulted in a reduction of deployment expenses related to ground stations and relay satellites, the multi-hop inter-satellite networks exhibit heightened susceptibility to nefarious attacks due to their primarily LoS propagation characteristics. The authors referenced in [5] also emphasized the necessity of considering the cumulative transmission latency of multi-hop inter-satellite networks within the framework of security concerns. A rogue node may potentially succeed in impersonating one of the legitimate nodes within ISL, thereby intentionally prolonging the data forwarding delay. Crisis situations may even exacerbate adverse outcomes, such as collisions and extensive devastation. The remedial approach proposed in [5] was predicated on the exploration of appropriate routing algorithms that leverage the understanding of the trustworthiness level of each satellite, in conjunction with other contemporary security measures, including encryption and digital signatures, among others. Nonetheless, the exploration of security challenges associated with mega-constellations remains in its nascent stages, thus necessitating immediate scholarly attention to address this critical gap.

In order to address the aforementioned challenges, we conduct an analysis of the prevailing advanced technologies and engage in a discourse regarding their prospective applications in satellite communication, encompassing the subsequent methodologies.

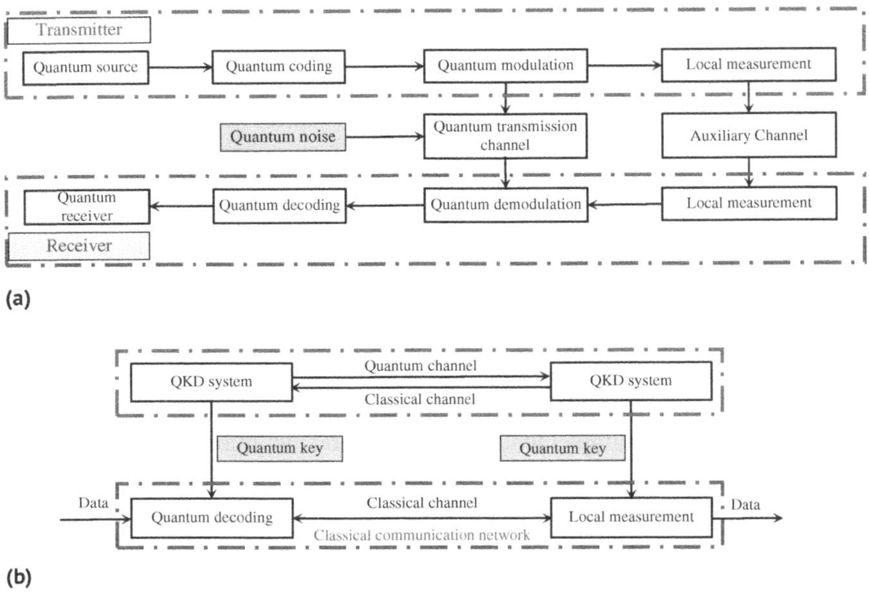

Figure 12.2 Schematic of the basic architecture of the mega-satellite constellation. (a) Basic model of quantum communication systems and (b) quantum secure communication system based on QKD.

12.2.1 Quantum Secure Communication

1. **Quantum Communication:** As illustrated in Figure 12.2, the fundamental framework of the quantum communication system alongside the QKD-based quantum secure communication system. Quantum communication exploits the non-clonability inherent in quantum states as well as the challenges associated with precise measurement, thereby facilitating secure information transfer via QKD [6]. Utilizing the Jinan-1 microsatellite as a case study, QKD technology is employed to synchronize the quantum key between the space station and the terrestrial station, thereby ensuring the integrity of information transfer. The merit of quantum communication is evidenced by its unconditional security, with any attempts at eavesdropping being detectable in real time. Nonetheless, quantum signals are vulnerable to disturbances caused by atmospheric turbulence and cosmic radiation, and the issues concerning the miniaturization, integration, and compatibility of equipment with preexisting systems require further investigation.
2. **Quantum Teleportation:** Quantum teleportation employs the phenomenon of quantum entanglement to facilitate the remote transmission of quantum

states. Entanglement refers to the condition in which two or more particles exhibit nonlocal correlations, such that the measurement of one particle instantaneously influences the state of another [7]. For the process of teleportation to occur, the sender and receiver must possess a shared pair of entangled particles before the sender conducts a measurement on one of them. Subsequently, the sender conveys the measurement outcome through a classical communication channel. The receiver then utilizes this information to execute a suitable unitary transformation on the entangled particle, thereby reconstructing the original quantum state. Despite the nonlocal characteristics exhibited by entanglement, the entirety of the teleportation process is ultimately governed by the limitations imposed by the speed of classical communication, thereby ensuring adherence to the principles of relativity.

The prospective applications of quantum teleportation in the domain of long-distance secure communication, particularly concerning deep-space communication, have attracted considerable scholarly interest. At present, this technology is confined to the realm of experimental research. The deep space optical communications laboratory at NASA has undertaken pertinent experiments to investigate the viability of quantum teleportation under conditions that simulate deep-space environments. These investigations concentrate on the optimization of entanglement distribution and the teleportation of quantum states by refining both experimental configurations and algorithms to improve transmission fidelity and success rates. Despite the practical implementation encountering significant hurdles, these research endeavors establish an essential technical groundwork for prospective secure deep-space communication predicated on quantum teleportation.

12.2.2 Laser Communication and Photonics

1. **Laser Communication:** As illustrated in Figure 12.3, the flowchart representing a prevalent laser communication system is utilized within the context of dual-satellite on-orbit communication links. Laser communication technology employs lasers to produce highly coherent light emissions, and through the modulation of parameters such as laser intensity, frequency, or phase, it facilitates the rapid encoding and transmission of information [8]. Given the short wavelength, pronounced directionality, and minimal divergence angle associated with lasers, this technology is adept at transmitting substantial volumes of information while exhibiting robust anti-jamming and anti-interception properties, thereby conferring notable advantages in terms of communication security and data transmission velocity. Nevertheless, the exceptionally high demands for alignment and tracking of laser links render

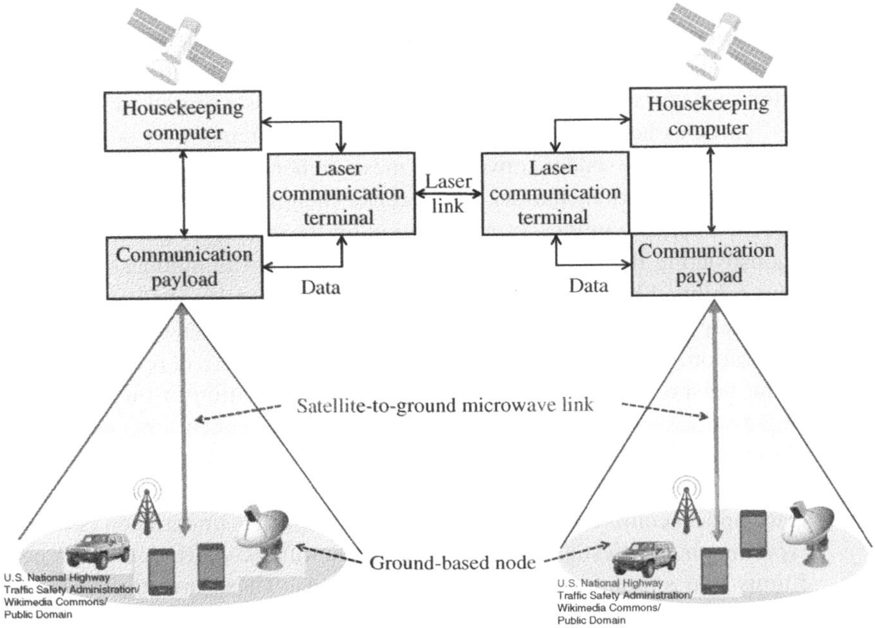

Figure 12.3 Schematic diagram of the on-orbit communication link of the laser communication binary satellite.

the practical realization of this technology a challenging endeavor concerning satellite attitude and orbital control.

2. **Photonic Integrated Circuit (PIC):** PIC has profoundly influenced the domain of optical communication systems [9]. This chip possesses the capability to perform a multitude of functions, which represents a significant advancement within this discipline. Concerning laser modulation, these PICs exhibit remarkable efficiency in the conversion of electrical signals to optical signals through the utilization of electro-optical modulators, which rapidly alter the intensity, phase, and frequency of light. The coding functionality is achieved via specialized optical configurations, such as wavelength coding through arrayed waveguide gratings, which facilitate the multiplexing and demultiplexing of multiple signals. Within the realm of quantum communications, the PIC can also incorporate a quantum key generation module that employs the principles of quantum optics to produce secure cryptographic keys. This elevated level of integration confers substantial advantages. The optical signals within the device traverse exceedingly short distances, thereby minimizing energy consumption and transmission losses.

Consequently, this enhances the device's portability and scalability, while simultaneously permitting seamless integration into miniaturized communication terminals.

The Phaeton initiative, an endeavor financed by DARPA, serves as a quintessential illustration of the implementation of PICs. Its objective is to engineer high-performance PICs intended for deployment on satellites and various other platforms in orbit. This initiative seeks to innovate state-of-the-art PICs for utilization in orbital environments. These semiconductor devices will amalgamate sophisticated optical communication functionalities, thereby minimizing power consumption and reducing the physical dimensions of satellites. The Phaeton circuit will facilitate rapid and dependable processing and transmission of data in the space environment, thereby offering significant technological advantages.

12.2.3 Design of Security Architecture for Cross-orbit Constellation

1. **Software-defined Networking (SDN) and Network Function Virtualization (NFV):** SDN and NFV technologies represent two groundbreaking innovations in network architecture that significantly enhance the flexibility and programmability of networks through the decoupling of the control plane from the data plane, alongside the implementation of centralized controllers to facilitate cohesive management and scheduling of the network. Concurrently, NFV permits an exceptionally adaptable configuration and real-time modification of network functions by virtualizing the capabilities traditionally associated with hardware devices, thereby deploying them as software on a standardized hardware infrastructure. Conversely, NFV allows for an exceptionally adaptable configuration and real-time modification of network functionalities through the virtualization of the roles of conventional hardware devices, which are deployed as software on a common hardware platform.

   NFV, conversely, facilitates a highly adaptable configuration and real-time modification of network functionalities by implementing the capabilities of conventional hardware apparatus in the guise of virtualized software on a unified hardware architecture. The synergistic application of SDN and NFV methodologies can considerably augment the flexibility, scalability, and responsiveness of space communication networks, recalibrating resource distribution and network topology by mission-specific criteria and instantaneous network conditions, while simultaneously promoting the seamless amalgamation and collaborative operation of inter-orbit communication networks [10]. Nevertheless, the application of these two technologies within the extraterrestrial milieu continues to encounter certain obstacles, particularly concerning software dependability, real-time operational efficacy, and compatibility with hardware infrastructures. Concurrently, the assurance of network security, stability, and safeguarding against malevolent intrusions and data breaches persist as critical challenges that necessitate resolution.

2. **Zero-trust Security Architecture:** The zero-trust security architecture is predicated on the foundational tenet of "never trust, always verify," necessitating rigorous authentication, authorization, and encryption for every access request and data transmission within a space communications network [11]. This technological framework adeptly mitigates security vulnerabilities and safeguards the integrity of the network and communication data, irrespective of the origin of the request, whether from internal or external networks. Notwithstanding its capacity to fortify the system's resilience against cyber-attacks, the implementation of a zero-trust architecture concurrently imposes elevated demands on computational and storage resources, thereby exacerbating the complexity and performance overhead associated with communications.
3. **Blockchain:** Blockchain employs various technologies, including distributed ledgers, cryptographic hash functions, and digital signatures, to facilitate decentralized data storage, interoperability, and traceability. In the realm of space communication, it has the potential to establish a secure communication framework and a robust data management system [12]. The primary advantage lies in the elimination of the necessity for a centralized trust authority, thereby allowing the involved parties to engage in direct secure data exchanges, mitigate trust-related risks, guarantee data integrity and authenticity, and facilitate straightforward auditing processes. Conversely, the primary challenge associated with blockchain technology is that its consensus mechanism demonstrates diminished efficiency within the space environment characterized by elevated latency and reduced throughput, thereby necessitating further investigation into lightweight blockchain solutions.

References

1 Pengfei Wang, Boya Di, and Lingyang Song. Mega-constellation design for integrated satellite-terrestrial networks for global seamless connectivity. *IEEE Wirel. Commun. Lett.*, 11(8):1669–1673, 2022.

2 Wenhao Liu, Jiazhi Wu, Quanwei Lin, et al. Efficient satellite-ground interconnection design for low-orbit mega-constellation topology. *IEEE Trans. Mob. Comput.*, 24(3):2098–2109, 2025.

3 Hai Yang, Bingli Guo, Kuan Yan, et al. Constellation structure design for LEO mega-constellation with optical intersatellite link. In *2022 Asia Communications and Photonics Conference (ACP)*, pages 1258–1261, 2022.

4 Inigo Del Portillo, Bruce G Cameron, and Edward F Crawley. A technical comparison of three low earth orbit satellite constellation systems to provide global broadband. *Acta Astronaut.*, 159:123–135, Jun. 2019.

5 Yafeng Zhan, Guanming Zeng, and Xiaohan Pan. Networked TT&C for mega satellite constellations: A security perspective. *China Commun.*, 19(9):58–76, Sep. 2022.

6 Jasminder S Sidhu, Siddarth K Joshi, Mustafa Gündoğan, et al. Advances in space quantum communications. *IET Quantum Commun.*, 2(4):182–217, 2021.

7 Jigang Ren, Ping Xu, Hailin Yong, et al. Ground-to-satellite quantum teleportation. *Nature*, 549(7670):70–73, 2017.

8 Morio Toyoshima. Recent trends in space laser communications for small satellites and constellations. *J. Lightwave Technol.*, 39(3):693–699, 2021.

9 Shannon Alt. Photonic integrated circuit (PIC) device structures: Background, fabrication ecosystem, relevance to space systems applications, and discussion of related radiation effects. Technical report,2016.

10 Michel Bonfim, Kelvin Dias, and Stenio Fernandes. Integrated NFV/SDN architectures: A systematic literature review. *ACM Comput. Surv.*, 51(6):1–39, 2019.

11 Yuanhang He, Daochao Huang, Lei Chen, Yi Ni, and Xiangjie Ma. A survey on zero trust architecture: Challenges and future trends. *Wirel. Commun. Mob. Com.*, 2022(1):6476274, 2022.

12 Mohamed Torky, Tarek Gaber, Essam Goda, Vaclav Snasel, and Aboul Ella Hassanien. A blockchain protocol for authenticating space communications between satellites constellations. *Aerospace*, 9(9):495, 2022.

13

Physical Layer Security for Satellite Communications

SCSs are progressively becoming essential components within contemporary networks, thereby rendering their security a paramount concern [1]. Owing to the inherently open character of space-based communication channels, these systems are susceptible to various threats, including eavesdropping, interception, and jamming. Although conventional security mechanisms such as encryption and authentication are frequently employed, they may not adequately safeguard against advanced attacks, particularly in volatile and adversarial settings. As illustrated in Figure 13.1, PLS has surfaced as an adjunctive solution, thereby enhancing the comprehensive security framework of SCSs [2].

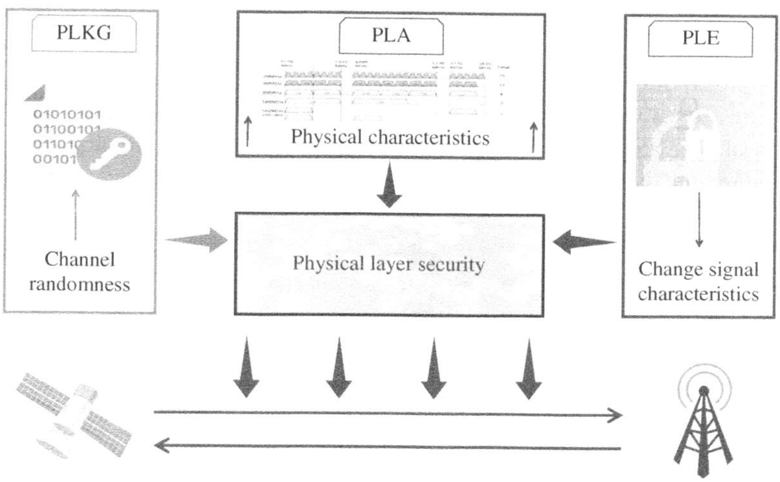

Figure 13.1 PLS for satellite communications.

Safeguarding Satellite Communications: Issues, Challenges, and Solutions, First Edition.
Jianping An, et al.

PLS capitalizes on the intrinsic properties of the physical communication medium, including phenomena such as signal attenuation, noise perturbations, and interference, to augment data confidentiality. It furnishes a formidable protective framework at the physical layer, functioning independently of higher-layer protocols, and serves to enhance their efficacy for augmented safeguarding. This chapter delves into three pivotal elements of PLS within satellite communications: physical layer key generation (PLKG), physical layer authentication (PLA), and physical layer encryption (PLE), each of which is indispensable for ensuring the confidentiality, integrity, and authenticity of communications between satellite and ground stations.

13.1 Physical Layer Key Generation

13.1.1 Description of PLKG

PLKG constitutes an essential security paradigm that leverages the distinctive properties of the communication channel to produce cryptographic keys at the physical layer. In contrast to conventional key generation techniques, which frequently depend on computational algorithms and protocols at higher layers, PLKG exploits stochastic fluctuations within the physical medium—such as channel noise, fading, and interference—to establish shared keys among authorized communicative entities. As illustrated in Figure 13.2, this methodology guarantees that the keys are produced in a manner that renders them challenging for potential adversaries to replicate, as they are intrinsically linked to the physical conditions of the channel at the moment of transmission.

PLKG presents numerous advantages in SCSs, notably its augmented security and resilience against eavesdropping attempts. Given that the cryptographic keys are generated dynamically in accordance with real-time channel conditions,

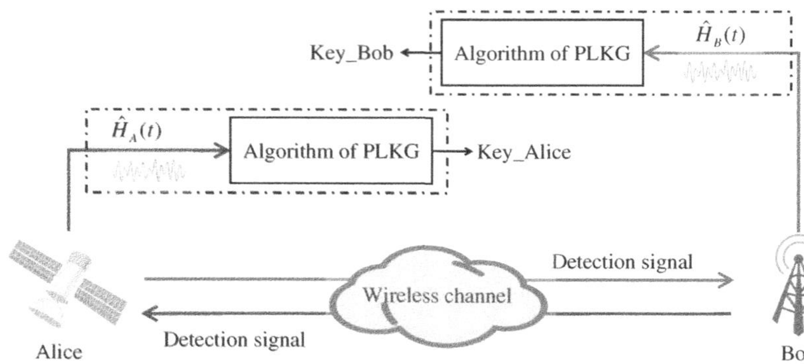

Figure 13.2 Principle of generating the physical layer key.

any potential adversary would encounter substantial obstacles in obtaining the identical key without being in close physical proximity to the communication link. Consequently, PLKG serves as an efficacious mechanism for facilitating secure communication between satellites and terrestrial stations, thereby providing an additional layer of defense against prospective threats within space-based communication networks.

13.1.2 Techniques of PLKG

The process of PLKG constitutes a methodology that exploits the intrinsic stochastic properties of wireless communication channels to derive cryptographic keys for the purpose of ensuring secure communication. This technique is predominantly classified into two distinct categories: source-based key generation and channel-based key generation.

13.1.2.1 Source-based Key Generation

The process of source-based key generation capitalizes on the intrinsic randomness inherent within the communication channel to facilitate the creation of secure shared keys. This methodology employs a variety of channel characteristics, including signal amplitude, phase, and noise, which are subject to influences from aspects such as multipath effects, fading, and interference that manifest during the transmission process. These stochastic fluctuations within the channel are pivotal for the formulation of a secure key, as they render the signals exceedingly challenging to anticipate or replicate by unauthorized entities. The reciprocity characteristic of the satellite communication channel guarantees that both parties engaged in communication receive congruent signals under optimal conditions, thereby allowing them to derive the same random source for the purpose of key generation.

The process typically involves several key steps [3]:

- **Shared Random Source Acquisition:** Both parties measure channel characteristics, such as amplitude and phase response, to obtain a shared random source. The reciprocity of the wireless channel ensures the consistency of this shared random source.
- **Quantization:** The shared random source is converted into a binary sequence. Similar to traditional analog-to-digital converters (ADCs), quantization transforms analog signals into digital form. In PLKG, the goal is to convert channel characteristics into binary data suitable for key generation.
- **Information Reconciliation:** The parties process the quantized data to correct discrepancies caused by channel noise and inconsistencies, ensuring the generated keys are consistent.
- **Privacy Amplification:** This technique enhances the security of the key, reducing potential eavesdropping risks.

Through adherence to these outlined procedures, PLKG facilitates the generation of secure shared keys that are predicated upon the distinctive attributes of the wireless communication channel.

13.1.2.2 Channel-based Key Generation

In contrast, the process of channel-based key generation emphasizes the statistical characteristics of the communication channel in order to derive a secure shared key. This approach capitalizes on the attributes of the channel, including phenomena such as fading, noise, and interference patterns, which are inherently stochastic and distinctive to the specific communication milieu. The parties engaged in communication exchange CSI—which encompasses measurements or estimations of various parameters such as signal strength, channel delay, and phase—thereby collaboratively constructing a confidential key. The primary advantage of this methodology resides in its capacity to produce keys grounded in the physical properties of the channel, which are exceedingly challenging for an external adversary to duplicate without having access to the authentic communication environment.

Each methodology presents a unique array of benefits and obstacles. Source-based key generation provides a high level of security owing to its dependence on the intrinsic randomness of the communication channel. Nevertheless, it is susceptible to channel noise, which can adversely influence the integrity of the generated key. Conversely, channel-based key generation may demonstrate greater resilience in real-world applications; however, it is not without its security vulnerabilities, particularly in scenarios where the CSI is compromised.

Scholars are diligently investigating hybrid methodologies that amalgamate components from both techniques to augment the security and efficacy of PLKG systems. Through the synthesis of the advantages inherent in source-based and channel-based key generation, these hybrid methodologies aspire to deliver more resilient and pragmatic solutions for secure communication within wireless networks.

13.1.3 Advantages and Challenges

13.1.3.1 Advantages

The PLKG is increasingly recognized within the domain of satellite communications as a means to enhance both security and operational efficacy. A primary advantage of PLKG lies in its capacity to utilize the intrinsic randomness present within the communication channel. Satellite communication channels are subject to various influences, including atmospheric phenomena, interference, and signal fading, which introduce a degree of unpredictability that can be harnessed for the generation of cryptographic keys. This inherent randomness complicates the efforts of potential adversaries to replicate the generated keys, thereby providing an additional layer of security. Furthermore, PLKG mitigates the dependence on centralized key management systems, which are frequently susceptible to interception, by facilitating the direct generation of unique cryptographic keys at the physical layer [4].

Another notable merit of PLKG in the domain of satellite communications pertains to its scalability and appropriateness for extensive networks. Satellite constellations, especially those operating in LEO, necessitate effective methodologies for key re-establishment in order to uphold secure communication over prolonged durations. PLKG provides a resolution that is readily adaptable to the dynamic and extensive characteristics inherent in satellite networks. Given that keys are generated in accordance with the physical attributes of the communication channel, they can be frequently and demandingly updated, thereby ensuring that keys remain both current and secure across all connections within the satellite network [5].

Moreover, PLKG exhibits considerable advantages regarding resource efficiency. In contrast to conventional cryptographic techniques, which often depend extensively on computational resources for key generation and encryption processes, PLKG leverages the inherent characteristics of the satellite communication channel itself. This diminishes the necessity for substantial computational power, rendering PLKG a compelling alternative for satellite systems characterized by limited processing capabilities or operational constraints in resource-scarce environments. This trait positions PLKG as an exemplary candidate for both commercial and defense satellite applications, where the preservation of secure and efficient communication is paramount [3]. By integrating high security with minimal overhead, PLKG emerges as a noteworthy trajectory for the advancement of satellite communication networks.

13.1.3.2 Challenges

In the domain of satellite communications, PLKG is paramount for the augmentation of security through the establishment of secure communication channels, eschewing a complete reliance on conventional cryptographic techniques. Nevertheless, the deployment of PLKG within satellite networks is beset by a myriad of challenges attributable to distinctive environmental and technical constraints. The considerable distances and inherently variable characteristics of satellite channels result in substantial fluctuations in both signal strength and quality, thereby complicating the key generation process. In contrast to more consistent terrestrial networks, satellite channels are susceptible to adverse weather phenomena, interference, and signal degradation, all of which can adversely affect the dependability of the key generation procedure [6]. Furthermore, the mobility of satellites, along with the potential presence of eavesdroppers—such as hostile satellites or ground-based stations—exacerbates the complexities surrounding the confidentiality and authenticity of the keys that are produced [4].

Furthermore, the scalability of PLKG within satellite systems constitutes a considerable impediment. As satellite constellations expand in both magnitude and intricacy, the challenge of facilitating secure key exchanges among each satellite and multiple receivers intensifies significantly. The discrepancies in orbital trajectories, the transition processes between satellites, and the diverse geographical positions of users exacerbate the synchronization necessary for effective key generation. Moreover, the deployment of secure PLKG methodologies must consider potential quantum vulnerabilities that may jeopardize conventional key generation techniques predicated on classical physics [7]. Consequently, proposed solutions must not only tackle these pragmatic obstacles but also foresee prospective threats, such as advancements in quantum computing, which could erode the efficacy of the existing cryptographic strategies utilized in SCSs. To guarantee that satellite networks can persist in delivering secure communication, continuous exploration into innovative methodologies, encompassing sophisticated signal processing techniques and quantum-resistant protocols, is imperative.

13.2 Physical Layer Authentication

13.2.1 Description of Physical Layer Authentication

PLA represents a sophisticated security paradigm that seeks to authenticate the identities of the communicating parties—such as satellites, terrestrial stations, or other relevant entities—through the assessment of the distinctive physical attributes inherent to the communication channel [8]. As illustrated in Figure 13.3, these attributes, which encompass signal power, phase, time-of-arrival (TOA), and Doppler shift, are inherently affected by the spatial position and dynamics of the involved entities. Consequently, these factors yield a robust and virtually unforgeable signature for each satellite or ground station.

In the domain of satellite communications, the PLA plays a crucial role in mitigating diverse forms of security threats, notably spoofing, wherein a nefarious actor masquerades as a legitimate satellite or terrestrial station to illicitly infiltrate the network. By leveraging the intrinsic characteristics of the physical layer that are intricately linked to the communication channel, adversaries are rendered incapable of replicating or fabricating these signals without detection. For example, satellites operating in LEO exhibit unique channel characteristics that exhibit dynamic variations attributable to their swift orbital motion. The TOA of the signal transmitted from a satellite to either a terrestrial station or another satellite is contingent upon the satellite's spatial orientation and kinematic state. Through the juxtaposition of these channel parameters with preestablished reference values or anticipated profiles, satellites and terrestrial stations are able to authenticate one another in real time [9]. This mechanism affords an augmented layer of security

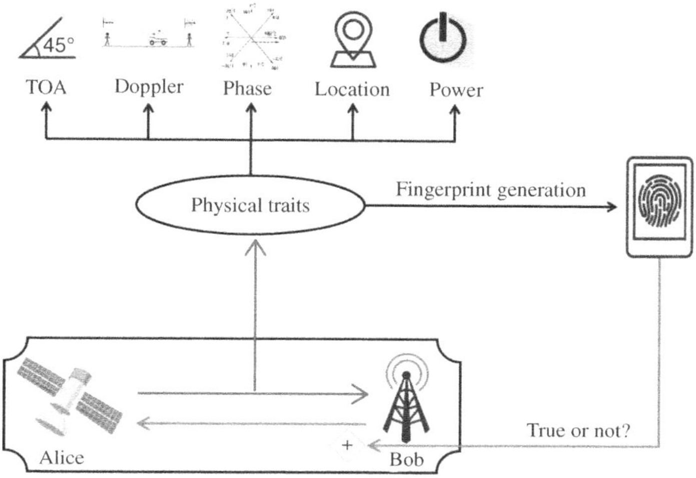

Figure 13.3 PLA in satellite communications.

that presents considerable challenges for adversaries to circumvent, particularly within the context of rapidly evolving satellite constellations.

However, the effectiveness of PLA can be limited by the presence of multipath interference, where signals reflect off obstacles and cause distortion in the channel. Additionally, large-scale networks with numerous satellites may require sophisticated algorithms to authenticate communication parties in a timely and efficient manner.

13.2.2 Techniques of PLA

A variety of methodologies are utilized for PLA, each contingent upon specific physical characteristics that can be leveraged for the purpose of authentication.

There exist two methodologies to ascertain the veracity of the signal by examining the variation in the physical distance between the satellite and the receiving station. One of the most prevalently employed methodologies is TOA-based authentication, which depends on the distinct temporal delay required for a signal to traverse from the satellite to the receiver. Given that this delay is intrinsically contingent upon the spatial coordinates of the satellite and the location of the ground station, it provides a dependable mechanism for authentication, thereby ensuring that only signals emanating from the anticipated satellite are acknowledged. Doppler shift–based authentication represents an additional frequently utilized approach. The Doppler effect manifests as a consequence of the relative motion existing between the satellite and the receiver, culminating in a shift in the frequency of the transmitted signal. This frequency alteration is distinctive to the satellite's trajectory, and through the analysis of the Doppler pattern, the receiver can effectively authenticate the satellite. This methodology is particularly advantageous owing to its capacity to yield a unique signature for each satellite contingent upon its motion.

Channel impulse response (CIR)-based authentication leverages the distinctive attributes of the communication channel. The CIR, which encapsulates the temporal and amplitude fluctuations of the signal as it traverses the environment, is subject to variations contingent upon factors such as the satellite's geographical position and the characteristics of the surrounding environment. Through the examination of the CIR, the receiver is capable of differentiating between legitimate and counterfeit signals, thereby rendering this methodology exceptionally resilient to spoofing endeavors. The phase of the transmitted signal additionally constitutes a significant parameter for the authentication of satellites. Authentication based on signal phase is affected by a myriad of factors, including the satellite's spatial coordinates, velocity, and prevailing environmental conditions. Consequently, the phase shift can yield a distinctive and dependable signature for the authentication of the satellite, thus establishing it as an efficacious instrument in the domain of communication security.

Another method of authentication is location-based authentication, which utilizes satellites' established orbital trajectories. Considering that each satellite adheres to a predictable path, its coordinates at any specific moment can be compared with the anticipated location. By juxtaposing the satellite's actual coordinates with the forecasted ones, the receiver is able to authenticate its legitimacy. This methodology guarantees that only signals emanating from a correctly situated satellite are deemed reliable.

Finally, the concept of transmission power–based authentication is also taken into account, as the transmission power of the satellite's signal is subject to fluctuations based on its orbital position and the prevailing channel conditions. Such variations prove challenging to duplicate, thereby providing an additional stratum of security against prospective adversaries endeavoring to replicate the satellite's transmission.

As illustrated in Table 13.1, various PLA methodologies present distinct benefits and drawbacks contingent upon their suitability and the specifications of the system.

13.2.3 Advantages and Challenges

13.2.3.1 Advantages

The implementation of PLA in SCSs presents numerous advantages, particularly in relation to the security imperatives of satellite networks. A primary advantage of PLA lies in its capacity to utilize the intrinsic attributes of the physical

Table 13.1 Comparison of PLA methods.

Authentication method	Advantages	Disadvantages	Applicability
TOA-based authentication	Simple, effective for satellite position-based authentication	Requires precise timing, susceptible to time-based attacks	Suitable for high-accuracy time-based satellite systems
Doppler shift-based authentication	Unique satellite signature, resistant to spoofing	Vulnerable to frequency-based attacks and noise	Best for systems with significant relative motion
CIR-based authentication	Highly resistant to spoofing due to signal properties	Complex, affected by interference or multipath	Best for secure communications with predictable environments
Signal phase-based authentication	Reliable through phase shift analysis	Requires precise phase measurement, sensitive to noise	Effective with high-precision phase measurements
Location-based authentication	Verifies satellite authenticity via location	Relies on precise location tracking, affected by orbital errors	Ideal for systems with predictable satellite trajectories
Transmission power-based authentication	Difficult to replicate due to power variations	Sensitive to environmental factors, difficult to measure	Works when signal power variations can be reliably measured

layer, such as channel randomness and signal characteristics, for the authentication of communication entities. This approach diminishes the dependence on conventional cryptographic frameworks, which often incur significant computational costs and are vulnerable to key distribution weaknesses. The distinctive propagation attributes of satellite channels, encompassing fluctuations in time, frequency, and phase, can be strategically harnessed to differentiate between authorized users and potential adversaries [10].

Moreover, PLA offers substantial resilience against a diverse array of cyber intrusions, such as spoofing and MITM attacks, as adversaries encounter considerable challenges in replicating the characteristics of the physical layer without detection. This inherent security, when integrated with methodologies such as TOA and angle-of-arrival measurement, markedly fortifies the overall security framework of SCSs.

Furthermore, the integration of ML algorithms with PLA has the potential to enhance detection precision and robustness, thereby rendering SCSs increasingly amenable to the dynamics of emerging threats [11].

13.2.3.2 Challenges

Despite its merits, the implementation of PLA in satellite communications encounters numerous challenges that impede its extensive adoption. A principal barrier is the intricate and dynamic characteristics inherent to the satellite communication environment. Satellites frequently function under highly fluctuating conditions, experiencing signal degradation due to atmospheric interference, precipitation-induced signal loss, and various other environmental influences. Such variations can induce alterations in the physical layer attributes, thereby complicating the authentication procedure. Moreover, the extensive coverage area of satellites, coupled with their elevated mobility, presents significant difficulties in sustaining consistent and precise measurements of signal characteristics. Consequently, the task of ensuring that the physical layer properties remain sufficiently stable for dependable authentication emerges as a considerable challenge. In addition, the constrained bandwidth present in SCSs exacerbates the complexity, as sophisticated PLA methodologies can demand substantial resources concerning data throughput, which is particularly crucial in low-bandwidth scenarios.

One additional challenge emerges from the escalating amalgamation of satellite networks with terrestrial infrastructures. The hybrid configuration of such systems engenders novel attack vectors, as nefarious entities may capitalize on the disparities between terrestrial and satellite-based physical layer attributes. The attainment of effective synchronization among these networks is paramount in mitigating vulnerabilities, thereby rendering the deployment of PLA across heterogeneous contexts a formidable endeavor. Finally, quantum computing presents a prospective threat to the integrity of SCSs, as quantum algorithms possess the capability to potentially circumvent existing PLA methodologies [12]. Consequently, continuous investigation is imperative to formulate quantum-resistant authentication frameworks that can guarantee sustained security within satellite networks.

13.3 Physical Layer Encryption

13.3.1 Description of PLE

The PLE utilizes the intrinsic characteristics of the communication channel to safeguard transmitted information from unauthorized interception by modifying the signal at the physical layer prior to its transmission. In contrast to conventional encryption methodologies, which function at upper layers such as the application or transport layers, PLE incorporates techniques including signal scrambling, BF, AI injection, and FH to render intercepted signals exceedingly challenging to decipher. This methodology markedly enhances the security of SCSs, particularly in scenarios where encryption at higher layers may prove insufficient due to computational constraints or the potential for advanced cyber-attacks. Consequently, even in the event that an adversary successfully intercepts the signal, they would be rendered incapable of extracting meaningful information absent the appropriate decryption apparatus.

While PLE provides augmented security measures, it concurrently poses challenges associated with its deployment. A primary obstacle is the necessity to achieve a harmonious equilibrium between security protocols and system performance [13]. The incorporation of artificial noise injection and the intricacies associated with BF may lead to increased energy consumption and a reduction in the data throughput of the communication system. Moreover, the dynamic and swiftly fluctuating channel conditions inherent in satellite communication networks necessitate the employment of adaptive methodologies capable of responding to the diverse characteristics of the communication medium.

13.3.2 Techniques of PLE

The PLE utilizes a variety of methodologies to manipulate the transmitted signal, thereby augmenting its security. BF-based encryption specifically directs the transmission beam toward a designated receiver, ensuring that the intended ground station is the sole recipient of the signal while concurrently thwarting unauthorized interception. This methodology can be synergistically integrated with artificial noise injection, wherein noise is superimposed onto the signal, thereby rendering it incomprehensible to potential eavesdroppers. Nevertheless, the legitimate recipient possesses the capability to effectively filter out this noise by leveraging known channel characteristics, thereby maintaining the integrity of the communication. Furthermore, transmit power control emerges as a crucial component in these strategies by facilitating dynamic adjustments to the transmission power. Such adjustments assist in constraining the signal's range, thereby diminishing the likelihood of interception by unauthorized entities and lessening interference, thereby enhancing the reliability of the communication.

In addition to the manipulation of signal transmissions, methodologies such as FH and MIMO-based encryption augment the complexity of communication, thereby rendering it more challenging for unauthorized entities to intercept the signal. FH entails the rapid alteration of the transmission frequency according to a predetermined or stochastic pattern, complicating the efforts of an eavesdropper to monitor the signal effectively. Conversely, MIMO-based encryption, which employs multiple antennas at both the transmitting and receiving ends, amplifies spatial diversity and obfuscates the signal, thereby elevating the difficulty for adversaries to derive meaningful information from the intercepted transmission.

Another significant methodology is physical-layer cryptography, which amalgamates conventional cryptographic paradigms with the distinctive attributes of the physical layer. This technique encodes the disseminated information in accordance with the characteristics of the transmission medium, such as channel response, thereby establishing an additional stratum of security. The synthesis of cryptographic methods and physical-layer strategies renders this approach particularly robust against adversarial attacks, thereby contributing another formidable line of defense against eavesdropping.

13.3.3 Advantages and Challenges

13.3.3.1 Advantages

PLE presents substantial advantages within the domain of satellite communications, specifically in terms of enhancing security while concurrently reducing computational complexity. A principal advantage of PLE lies in its capacity to secure communications directly at the physical layer, wherein encryption keys are derived from the intrinsic properties of the communication channel itself. This diminishes dependence on conventional cryptographic key distribution systems, which are susceptible to interception or compromise [2]. PLE capitalizes on the inherent randomness within the transmission channel, encompassing noise and interference, to obfuscate the transmitted data, thus delivering robust protection against eavesdropping and unauthorized interception. This attribute is particularly advantageous in satellite communication networks, where secure transmission across extensive distances and under variable conditions is paramount. Given that the encryption mechanism is intrinsically linked to the physical characteristics of the communication channel, it exhibits considerably greater resilience to attacks in contrast to traditional encryption techniques that emphasize upper-layer encryption and key management.

An additional advantage of PLE is its superior resource efficiency compared to conventional encryption methodologies. SCSs, particularly those functioning in LEO, frequently encounter limitations regarding power, bandwidth, and processing capabilities. The encryption mechanism employed by PLE, which does not necessitate extensive computational resources, aligns effectively with these operational constraints. It presents a streamlined encryption strategy that preserves both the performance and energy efficiency of SCSs, rendering it particularly suitable for next-generation networks where the optimization of resource utilization is paramount. Moreover, PLE can be integrated effortlessly with preexisting cryptographic techniques, thereby providing an additional security layer while simultaneously minimizing overhead and enhancing scalability within large-scale satellite constellations.

13.3.3.2 Challenges

Notwithstanding the benefits, the application of PLE within SCSs presents numerous obstacles. A primary concern is the intrinsic unpredictability associated with the communication channel in space-based systems. Satellite communication links are influenced by a myriad of environmental variables, including precipitation-induced signal loss, disruptions in the ionosphere, and variations in atmospheric pressure, all of which can result in considerable fluctuations in both signal integrity and strength [14]. These environmental variables contribute to the intricacy of the encryption mechanism, as the stochastic elements utilized for the generation of encryption keys may not consistently exhibit stability or predictability. Furthermore, ensuring dependable encryption within LEO satellite constellations, characterized by frequent transitions among satellites, presents particular difficulties. The synchronization and management of encryption keys across diverse satellite links necessitate ongoing updates to uphold a secure communication framework, thereby augmenting the overall complexity of the system.

Another formidable challenge resides in the amalgamation of PLE with extant higher-layer cryptographic frameworks. Although PLE furnishes robust security at the physical stratum, it cannot supplant conventional cryptographic methodologies for E2EE. The implementation of hybrid systems that integrate both PLE and higher-layer cryptographic strategies is imperative to guarantee holistic security. Nonetheless, the orchestration between these disparate layers may engender additional overhead and augment the complexity of the system, particularly when endeavoring to sustain compatibility with legacy satellite communication infrastructures. Moreover, the swift progression of quantum computing constitutes a burgeoning threat to traditional encryption paradigms. Scholars are investigating methodologies to enhance the resilience of PLE against quantum-centric assaults, necessitating novel strategies for key generation and encryption frameworks. Safeguarding SCSs against the perils posed by quantum computing will demand persistent innovation in both PLE and post-quantum cryptographic algorithms.

References

1 Pingyue Yue, Jianping An, Jiankang Zhang, et al. Low earth orbit satellite security and reliability: Issues, solutions, and the road ahead. *IEEE Commun. Surv. Tutor.*, 25(3):1604–1652, 2023.

2 Nora Abdelsalam, Saif Al-Kuwari, and Aiman Erbad. Physical layer security in satellite communication: State-of-the-art and open problems. *IET Commun.*, 19(1):e12830, 2025.

3 Bin Li, Zesong Fei, Caiqiu Zhou, and Yan Zhang. Physical-layer security in space information networks: A survey. *IEEE Internet Things J.*, 7(1):33–52, 2020.

4 Pietro Tedeschi, Savio Sciancalepore, and Roberto Di Pietro. Satellite-based communications security: A survey of threats, solutions, and research challenges. *Comput. Netw.*, 216: 109246, 2022.

5 Radhika Radhakrishnan, William W Edmonson, Fatemeh Afghah, Ramon Martinez Rodriguez-Osorio, Frank Pinto, and Scott C Burleigh. Survey of inter-satellite communication for small satellite systems: Physical layer to network layer view. *IEEE Commun. Surv. Tutor.*, 18(4):2442–2473, 2016.

6 Qingjiang Xiao, Jinrong Zhao, Sheng Feng, Guyue Li, and Aiqun Hu. Securing NextG networks with physical-layer key generation: A survey. *Secur. Saf.*, 3:2023021, 2024.

7 Yongpeng Wu, Ashish Khisti, Chengshan Xiao, Giuseppe Caire, Kai-Kit Wong, and Xiqi Gao. A survey of physical layer security techniques for 5G wireless networks and challenges ahead. *IEEE J. Sel. Areas Commun.*, 36(4):679–695, 2018.

8 Mohammed Abdrabou and T Aaron Gulliver. Physical layer authentication for satellite communication systems using machine learning. *IEEE Open J. Commun. Soc.*, 3:2380–2389, 2022.

9 Gabriele Oligeri, Savio Sciancalepore, Simone Raponi, and Roberto Di Pietro. PAST-AI: Physical-layer authentication of satellite transmitters via deep learning. *IEEE Trans. Inf. Forensics Secur.*, 18:274–289, 2023.

10 Anil Kumar Kamboj, Poonam Jindal, and Pankaj Verma. Machine learning-based physical layer security: Techniques, open challenges, and applications. *Wirel. Netw.*, 27:5351–5383, 2021.

11 Lamia Alhoraibi, Daniyal Alghazzawi, Reemah Alhebshi, and Osama Bassam J Rabie. Physical layer authentication in wireless networks-based machine learning approaches. *Sensors*, 23(4):1814, 2023.

12 Abraham Sanenga, Galefang Allycan Mapunda, Tshepiso Merapelo Ludo Jacob, Leatile Marata, Bokamoso Basutli, and Joseph Monamati Chuma. An overview of key technologies in physical layer security. *Entropy*, 22(11):1261, 2020.

13 Suhyeon Jeon, Jeongho Kwak, and Jihwan P Choi. Cross-layer encryption of CFB-AES-TURBO for advanced satellite data transmission security. *IEEE Trans. Aerosp. Electron. Syst.*, 58(3):2192–2205, 2022.

14 Shuai Han, Jixi Li, Weixiao Meng, Mohsen Guizani, and Siyue Sun. Challenges of physical layer security in a satellite-terrestrial network. *IEEE Netw.*, 36(3):98–104, May 2022.

Index

Safeguarding Satellite Communications: Issues, Challenges, and Solutions, First Edition.
Jianping An, et al.

d

e

q

r

s

t

Printed and bound by CPI Group (UK) Ltd, Croydon, CR0 4YY
17/09/2026
14964232-0006